RHEOMETERS
FOR MOLTEN
PLASTICS

RHEOMETERS
FOR MOLTEN
PLASTICS

A PRACTICAL GUIDE TO
TESTING AND PROPERTY
MEASUREMENT

John M. Dealy

Department of Chemical Engineering
McGill University
Montreal, Canada

 Sponsored by the Society of Plastics Engineers

VAN NOSTRAND REINHOLD COMPANY
NEW YORK CINCINNATI TORONTO LONDON MELBOURNE

6654-0604

CHEMISTRY

Van Nostrand Reinhold Company Regional Offices:
New York   Cincinnati

Van Nostrand Reinhold Company International Offices:
London   Toronto   Melbourne

Copyright © 1982 by Van Nostrand Reinhold Company

Library of Congress Catalog Card Number: 81-7553
ISBN: 0-442-21874-5

Manufactured in the United States of America

Published by Van Nostrand Reinhold Company
135 West 50th Street, New York, N.Y. 10020

Published simultaneously in Canada by Van Nostrand Reinhold Ltd.

15 14 13 12 11 10 9 8 7 6 5 4 3 2 1

Library of Congress Cataloging in Publication Data

Dealy, John M.
   Rheometers for molten plastics

   (A Society of Plastics Engineers technical
monograph)
   Bibliography: p.
   Includes index.
   1. Plastics—Testing.   2. Rheology.   3. Rheo-
meters.   I. Title.   II. Series: Society of Plastics
Engineers technical monograph.
TA455.P5F29   668.4'197   81-7553
ISBN 0-442-21874-5   AACR2

To Jacqueline

Foreword

The Society of Plastics Engineers is pleased to sponsor and endorse this new publication, *Rheometers for Molten Plastics*. The volume serves a long-standing need and represents yet another joint contribution by the publisher and the Society to the technical literature of plastics science.

This publication should be particularly useful to the plastics engineer with a need to evaluate resin processability on the basis of rheological tests. The mathematics are unsophisticated and formulas are presented rather than derived. Further, emphasis on techniques of conversion of experimental data into meaningful property values ensures that the real need of the practicing engineer is served by this volume.

SPE has long sponsored books through the medium of its Technical Volumes Committee which has made a particular effort to encourage publication in areas of the technology not previously covered in the literature. The Committee has commissioned and directed such publications and reviewed final manuscripts for technical accuracy.

In addition to educational programs, conferences and meetings the Society publishes four periodicals—*Plastics Engineering, Polymer Engineering and Science (PE & S), Journal of Vinyl Technology* and *Polymer Composites*—as well as conference proceedings and other selected publications. Information on these activities can be obtained by contacting the Society of Plastics Engineers, 14 Fairfield Drive, Brookfield Center, Connecticut 06805.

ROBERT D. FORGER
Executive Director
Society of Plastics Engineers

Technical Volumes Committee

James L. Throne, Chairman
Lee L. Blyler, Jr.
A. Wayne Cagle
James Corbin
Jerome L. Dunn
Tom W. Haas

John L. Kardos
Richard R. Kraybill
Timothy Lim
Eldridge M. Mount, III
Nikolaos A. Peppas

Preface

This book was written for everyone who deals with polymers in the liquid state, whether they be quality control technicians, plastics engineers or research scientists. It contains the information necessary to choose the best rheological test for a given application and to build or purchase the instrument best suited to carrying out this test.

Since molten polymers are viscoelastic fluids, the measurement of a single rheological property does not provide a complete characterization. For example, it is not unusual to find that two resins have similar viscosity curves but differ markedly when compared on the basis of some other property (such as the extensional viscosity). For the plastics engineer, this means that each property provides a different measure of processability, and in the use of rheological tests for quality control or for the evaluation of a series of resins for possible use in a particular process, it is essential to select the test appropriate to the process of interest.

Many "horror stories" can be told that involve failure to use rheology to advantage: The processor who went bankrupt as a result of his failure to employ even the simplest quality control test to detect batch to batch variations in resin processability; the carloads of resin returned to a supplier because of a contract specification involving a particular rheological test that had little relationship to the aspects of melt behavior of actual importance in the process in which the resin was being used; the elaborate and costly laboratory evaluation program that was totally unsuccessful in identifying resins likely to be suitable for use with a particular forming machine.

The search for correspondences between rheological properties and various aspects of melt behavior in commercial processes is an active area of research, and much remains to be learned. However, there has been a rapid development in the past ten years of new rheometrical techniques and new commercially manufactured test instruments, so that one can choose from among a much wider range of tests, and this greatly increases the chances of identifying properties that are closely related to processing behavior.

It is intended that the largest part of the book be accessible to readers without a sophisticated mathematical background. No knowledge of tensors is assumed, and the intelligent use of most of the rheometrical formulas requires only elementary algebra. The emphasis is on presenting formulas and explaining their limitations, rather than on deriving them.

The chapters can be grouped into three sections. The first, consisting of Chapters 1 through 3, provides background information necessary to understand the later material on experimental methods and is provided for those readers who have had no formal training in rheology. Chapters 4 through 7 provide the basic rheometrical equations necessary to convert raw experimental data into meaningful property values; these chapters also describe a number of melt rheometers discussed in the technical literature. Finally, Chapters 8 through 11 contain information to help the reader to identify those tests and commercial instruments most likely to be useful in a particular application.

The information on commercial testers and rheometers is based on manufacturers' specifications, and every effort was made to include every relevant instrument, regardless of the country of manufacture. If any have been omitted, it is a result of oversight rather than deliberate omission. It was not possible to include all the information and photos supplied by some manufacturers, but I am nonetheless grateful to those few who sent full information in response to my first request. As for the rest, a barrage of telephone calls, cables and telex messages was ultimately successful in eliciting essential data on every instrument thought to be of use in the study of molten polymers. Addresses of every manufacturer mentioned in this book are listed in Appendix B.

Finally, I have the happy task of thanking some of the many who helped me to accomplish this project. I am grateful to McGill University, especially to W. J. M. Douglas, for making it possible for me to devote an extended period of full-time effort to the writing. Likewise, I owe a great debt to the University of Delaware, especially to A. B. Metzner, for providing an excellent environment in which most of the work was completed. It is not feasible to list all those who have provided useful information and who have stimulated my interest in the subject, but notable among these are F. N. Cogswell, J.-M Charrier, M. M. Denn, C. D. Denson, J. Goddard, C. D. Han, M. R. Kamal, H. M.

Laun, A. S. Lodge, C. W. Macosko, H. Münstedt, J. R. A. Pearson, C. J. S. Petrie, G. V. Vinogradov, J. L. White and K. Wissbrun. Special thanks are due to Chris Macosko, who read the entire manuscript and made many helpful suggestions. As for my firsthand knowledge of the problems involved in the design and use of melt rheometers, this has all come my way through the agency of my able and long-suffering research student-collaborators.

The Technical Volumes Committee of the Society of Plastics Engineers encouraged me to undertake the writing of this book and provided some useful guidance as to content.

The manuscript was typed with exceptional care and skill by Pat Fong. Most of the drawings were prepared by Ursula Seidenfuss, and several others were supplied by Jean Leblanc.

John M. Dealy

CONTENTS

Chapter 1
Introduction

1.1 WHY WE MEASURE RHEOLOGICAL PROPERTIES

Rheological properties of molten polymers are of practical interest to all those sectors of the plastics industry in which materials are processed in the molten state. This includes resin manufacturers and compounders, machinery manufacturers, and plastics processors. The uses to which property data are put include the evaluation of experimental resins, the selection of a resin for a particular process, quality control, screening of experimental resins, and process modeling.

If a processor wishes to check for uniformity in batches of a given resin, a simple measurement of melt index may suffice. On the other hand, if one wishes to select, from among a group of resins, the resin which will give the best performance in a particular process, it will probably be necessary to measure several properties. Finally, if one wishes to derive a mathematical model of a process, a more complete characterization of the melt may be needed, depending on the complexity of the process. And, of course, it should be remembered that in the modeling of a complete process, right up to the formation of the finished product, properties other than rheological ones are likely to be important. These might include heat capacity, thermal conductivity and density, all as functions of temperature and pressure. In summary, we can say that the specific choice of property to be measured will depend on:

1. The general chemical structure of the resin.
2. The process in which the resin is to be used.
3. The use to which the data are to be put.

Before proceeding to discuss some specific rheological properties and why it is that polymeric liquids exhibit complex rheological behavior, it is necessary to ensure that the reader has an adequate understanding of some basic concepts regarding the mechanical behavior of materials. The next few sections are intended to provide the basis for such an understanding.

1.2 SOME FUNDAMENTAL ASSUMPTIONS

In designing rheological experiments and interpreting the resulting data, certain assumptions are usually made. These assumptions are generally accepted as valid for molten polymers, and most of the equations presented in this book are derived by use of them. It is important to recognize this so that one can be alert to the exceptional situations in which one or more of them may not be valid.

First, we usually assume the melt to be incompressible so that its density is uniform throughout the field of flow of a rheometer. Of course, it is true that density depends on temperature and pressure, so that the validity of this assumption is related to the extent to which these quantities are constant throughout the melt. Since we normally take pains to maintain a reasonably uniform temperature in the sample, nonuniformity of density is likely to be a problem only in flows that involve large pressure gradients; for example, in a capillary rheometer operated with a large driving force.

We will use the symbols v and ρ to indicate the velocity and density at a point in the melt. Now, in fact, if we look at the melt from a microscopic point of view, we know that it consists of molecules, and that these molecules do not fill all the available space. If we focused attention on a point in the "free volume," we would see only empty space, while if we focused attention on one of the points occupied by a molecule, we would see a local motion representing the resultant of the average bulk flow of the material and the local random movement associated with Brownian motion. The use of a quantity such as velocity to describe the state of motion at a "point" in a material involves the assumption that the material can be considered, for most practical purposes, to be a "continuum" rather than a collection of discrete subunits. The validity of this assumption derives from the fact that over the smallest distance scale likely to be of any practical significance, there will always be such a large number of molecules that it is possible to identify a meaningful average of density and velocity.

A third assumption that will be made throughout the book is that the material of interest is "isotropic in its rest state." This means that the properties of the material have no intrinsic dependence on direction. Wood and graphite are examples of materials that are highly anisotropic. Polymeric liquids nearly always have anisotropy induced in them when they are deformed, but when allowed to stand free of deforming stresses for a sufficient length of time, this anisotropy will usually disappear. The only known exceptions to this general behavior of polymer melts are "liquid crystal" systems, such as certain polyesters.

1.3 INDEX NOTATION

In the presentation of fundamentals in the first two chapters of this book, it will be convenient to make use of index notation to refer to coordinate axes, velocity components, and stress components. The three spatial variables used to specify a point in rectangular coordinates will be referred to as x_1, x_2 and x_3. The same indexes will also be used with vector components. For example, the velocity in the direction of the x_2 coordinate axis will be designated v_2. Some of the quantities we will make use of—among them, stress and rate of deformation—have components requiring two subscripts for their specification; for example, τ_{13}. The physical significance of the nine possible combinations of subscripts of such quantities will be given later in this chapter.

It will frequently prove useful to refer to a typical spatial variable or component by use of a letter index (for example: x_i, v_j, τ_{ij}). The use of this nonspecific index implies that the statement or equation in which it is used is equally valid with the index set equal to any one of its three possible values.

1.4 STRAIN AND STRESS

Rheology is the study of those properties of materials that govern the relationship between stress and strain. For example, the rheological properties of an elastic solid govern the stress required to stretch a sample to a certain extent. In order to use the relationships presented in this book for calculating rheological properties from experimental data, it is necessary to have a good understanding of the terms stress and strain, and this section is provided for those readers who may be uncertain about the way these terms are used in rheology.

First, let's look at strain, which is a quantitative measure of the extent to which an element of material has been deformed. We note that defor-

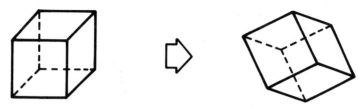

Figure 1-1. Rotation without deformation.

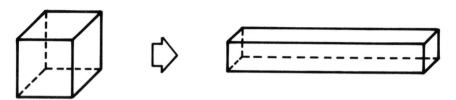

Figure 1-2. Uniaxial extensional deformation.

mation implies a change of shape. If you pick up an orange and turn it around to look for blemishes, you will rotate but not deform it, so there is no strain. Rotation without deformation is illustrated in Figure 1-1. If, however, you squeeze it to see how firm it is, you will deform it. Firmness is a rheological property, although a poorly defined one, and we can state generally that it is necessary to deform a material to learn anything about its rheological properties. To put it another way, you will learn nothing about the firmness of the orange if you rotate it without squeezing it.

There are many ways of defining quantitative measures of strain. For example, consider the elongation of a sample as shown in Figure 1-2. We could simply take the change in length as a measure of strain:

$$\text{Strain} \equiv L - L_0.$$

This is not a useful measure, however, because the amount of strain depends on the initial length. We would like to find a definition such that, if the sample is uniformly deformed, each element of the sample undergoes exactly the same strain. An obvious way of doing this is to define the strain as

$$\text{strain} \equiv \frac{L - L_0}{L_0} = \frac{L}{L_0} - 1.$$

This is called the Cauchy strain. Another quantity that is zero when the material is unstretched ($L = L_0$) is the Hencky strain, defined as

$$\epsilon \equiv \ln(L/L_0). \tag{1-1}$$

As we will see later, this latter measure of extensional strain is the one that proves most convenient for use in defining those rheological properties of liquids based on stretching deformations.

Another type of deformation that is very important in rheology is simple shear, which is illustrated in Figure 1-3. This is the type of deformation generated when a material fills the gap between two parallel, flat plates, and the upper plate is displaced linearly through a distance X_w, as illustrated in Figure 1-4.

If there is no slip at the surface of the plates, and if inertia can be neglected, each element of the material will be subjected to exactly the same local deformation. A quantitative measure of this deformation is the shear strain, γ, defined in Equation 1-2.

$$\gamma = \delta X_1/\delta x_2 \tag{1-2}$$

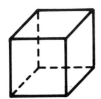 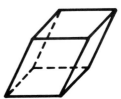

Figure 1-3. Simple shear deformation.

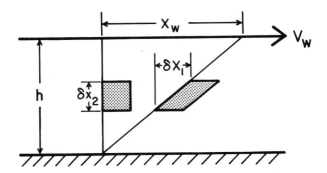

Figure 1-4. Shearing between parallel plates.

As can be seen in Figure 1-4, δx_2 is the height of the element of material, and δX_1 is the displacement of the upper surface of the element in the x_1 direction relative to its lower surface. Since the deformation is uniform, the strain is independent of the size of the element and can thus be related to the quantities h and X_w as follows.

$$\gamma = X_w/h \tag{1-3}$$

For the study of fluids, we often wish to subject the material to continuous shearing at a constant rate. This can be accomplished by moving the upper plate in Figure 1-4 at a constant velocity, V_w, in the x_1 direction. The fluid velocity at any point is given by Equation 1-4.

$$v_1 = \frac{x_2}{h} V_w \tag{1-4}$$

This flow is called "steady simple shear" and the local "shear rate" is defined as

$$\dot{\gamma} \equiv \frac{d\gamma}{dt} = \frac{dv_1}{dx_2}. \tag{1-5}$$

The shear rate (also called the rate of deformation or the strain rate) has units of reciprocal time, and is always given in s^{-1} (reciprocal seconds). Using Equation 1-4, the shear rate can be related to the quantities h and V_w as in Equation 1-6.

$$\dot{\gamma} = V_w/h \tag{1-6}$$

Note that V_w and h can be measured without disturbing the flow. Thus, in principle, it is a simple matter to subject a material to a uniform shearing strain with a known and easily controlled shear rate, and this is one of the reasons this flow is of great importance in rheology. Of course, to produce a completely uniform simple shearing motion, it is necessary that the two parallel plates be of infinite extent, and this is not practical for use in a finite laboratory. Methods for approximating simple shear in the laboratory will be described in Chapters 4 and 5.

We turn now to the definition of stress in rheology. A stress is a force per unit area. In S.I. units,* the stress is given in newtons per square meter (N/m^2). Because of its importance in mechanics, this combination of units has been given its own name, the pascal, and symbol (Pa). To describe the state of stress in the neighborhood of a point in a material, it is necessary, in general, to specify nine quantities. To see why this is so, refer to Figure 1-5, where there is depicted a small element of material at the point of interest.

We note that two types of stress can act on this element. "Normal stresses" act in a direction perpendicular to a face, and "shear stresses" act in a direction parallel to a face. Thus, at any given instant, stresses are acting in three directions on faces lying in three perpendicular planes, so that there are altogether nine components of stress. These are referred to by the use of a doubly subscripted symbol (σ_{ij}) where the first subscript refers to the orientation of the face upon which the force acts, and the second subscript indicates the direction of the force. We use the coordinate axis index, i, to indicate the orientation of a face, this index referring to the axis perpendicular to that face. For example, the symbol, σ_{21}, represents the stress acting on a face perpendicular to the x_2 coordinate axis with the force acting in the direction of the x_1 coordinate axis.**

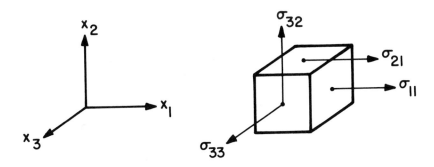

Figure 1-5. Some typical stress components.

---

*Appendix A contains a summary of definitions and conversion factors for units of quantities of interest in rheology.
**The set of quantities we have called the stress components obeys certain relationships which place it in a mathematical category of entities called "tensors." As a result, many powerful theorems in tensor algebra and calculus can be used to solve boundary value problems involving the deformation of continua. However, since this book is not concerned with solving boundary value problems, no further reference to the tensorial nature of the stress will be made.

To complete our definition of stress, we need to establish a sign convention. In this book, the following convention will be used: Stress is positive when acting in the positive direction of the coordinate axis indicated by the second subscript and on the face of the element having the largest value of x_i, where i is the first subscript.* Thus, all the stress components shown in Figure 1-5 are positive when acting in the direction of their respective arrows. Also, we note that normal stresses are positive when they are tending to stretch the element (tensile stress) and negative when they are tending to compress it (compressive stress).

It is customary to write out the components of the stress in the form of a matrix as follows.

$$\begin{bmatrix} \sigma_{11} & \sigma_{12} & \sigma_{13} \\ \sigma_{21} & \sigma_{22} & \sigma_{23} \\ \sigma_{31} & \sigma_{32} & \sigma_{33} \end{bmatrix}$$

The principle of moment of momentum, one of the basic laws of mechanics, can be used to prove that the stress matrix is symmetrical, so that

$$\sigma_{ij} = \sigma_{ji} \tag{1-7}$$

for any values of i and j. For example, $\sigma_{13} = \sigma_{31}$. As a result of this, we can see that there are really only six independent stress components in place of nine.

There is one further aspect of the use of the concept of stress in rheology. This is, to put it as concisely as possible, that the absolute magnitude of the normal stress is not rheologically meaningful if the material of interest is incompressible. Only shear stresses and differences between normal stresses acting in different directions have rheological significance. To understand this, it may be helpful to return to the problem of determining the firmness of an orange. If the orange is incompressible (i.e., if its total volume is assumed unchangeable by altering the external pressure), we will not be able to deform it by changing the external pressure (for example, by placing it in a hyperbaric chamber). Thus, we will learn nothing about its rheological properties as long as

---

*The opposite convention is used in some of the literature on transport phenomena and kinetic theory, specifically in the important books by R. B. Bird *et al.*, but the one used here is traditional in mechanics and rheology.

only balanced normal stresses are applied. Only if we apply a greater stress in one direction than in another will we observe a deformation and find out something about the orange.

In terms of the stress at a point in a material, what we have said is that a quantity such as σ_{11} has no rheological significance, whereas $\sigma_{11} - \sigma_{22}$ does. To remind ourselves of this, it is customary to use, in place of the actual total stress, σ_{ij}, another quantity, called the "viscous stress" or the "extra stress," τ_{ij}, which differs from the stress by an arbitrary isotropic stress. An isotropic stress is one which has no shear components and in which the three normal components are equal, as shown in the matrix below.*

$$\begin{bmatrix} \pi & 0 & 0 \\ 0 & \pi & 0 \\ 0 & 0 & \pi \end{bmatrix}$$

Thus, the extra stress is defined by letting it be an anisotropic contribution to the total stress.

$$\sigma_{ij} = \tau_{ij} + \pi\delta_{ij} \tag{1-8}$$

where

$$\delta_{ij} = 1 \quad \text{when} \quad i = j$$
$$\delta_{ij} = 0 \quad \text{when} \quad i \neq j$$

Since π is unspecified, τ_{11} has no absolute significance, but note that

$$\sigma_{11} - \sigma_{22} = \tau_{11} - \tau_{22}$$
$$\sigma_{22} - \sigma_{33} = \tau_{22} - \tau_{33}$$
$$\sigma_{12} = \tau_{12}$$
$$\sigma_{13} = \tau_{13}$$
$$\sigma_{23} = \tau_{23}$$

---

*In a fluid at rest, the stress is isotropic and the pressure is equal to the negative of the quantity we have called π. However, the concept of pressure has no absolute significance in the continuum mechanics of non-Newtonian fluids, and the use of the word "pressure" in describing the stress in a deforming fluid can be misleading.

Thus, *shear* stresses and normal stress *differences* are unaffected by the use of the extra stress in place of the total stress, so that the arbitrariness of π does not introduce any uncertainty into the values of these quantities.

1.5 THE NEWTONIAN FLUID

A fundamental rheological property of a fluid is its viscosity. This is defined as the ratio of the shear stress to the shear rate in a simple shear flow.

$$\eta \equiv \tau_{21}/\dot{\gamma} \qquad (1\text{-}9)$$

For single phase liquids containing only low molecular weight compounds, the viscosity is independent of the shear rate and depends only on the temperature and pressure.

$$\eta = \eta(T, P) \qquad (1\text{-}10)$$

In fact, the rheological properties of all materials depend on temperature and pressure, and having stated this, we will not continue to list T and P as independent variables. Thus, for the purposes of this discussion, we will say that the viscosity of a Newtonian liquid is "constant," because it does not depend on the dynamical independent variable, $\dot{\gamma}$.

In fact, for low molecular weight, single phase liquids, this single "material constant," the viscosity, is sufficient to predict the response of the liquid to any type of deformation. The general relationship between stress and velocity gradients for such materials is given by Equation 1-11.

$$\tau_{ij} = \eta \left(\frac{\partial v_i}{\partial x_j} + \frac{\partial v_j}{\partial x_i} \right) \qquad (1\text{-}11)$$

The quantity in brackets on the right is called the "rate of deformation." Like the stress, it has nine components, only six of which are independent. (It is like the stress also in that it has certain mathematical properties which are said to define a tensor.)

Equation 1-11 is an example of a "rheological equation of state" or "constitutive equation," because it tells us everything we need to know about the material in order to construct differential equations which can be used, in principle, to predict the deformation arising from any type of external strains or stresses.

A material which obeys this particular constitutive equation is called a Newtonian fluid. Thus, we can say that single phase, low molecular weight liquids are Newtonian fluids, and once we have measured the viscosity of such a material, at any single shear rate, we have a complete rheological characterization of it (at a particular temperature and pressure). Molten polymers are decidedly non-Newtonian (i.e., they do not obey Equation 1-11), but it is still of interest to compare the behavior of melts to that of a Newtonian fluid, and we will have frequent occasion to do this in Chapter 2.

1.6 NON-NEWTONIAN PHENOMENA

Emulsions, suspensions of solids, polymer solutions and polymer melts are usually "non-Newtonian fluids." Thus, they do not obey Equation 1-11, and a single measurement of viscosity is not sufficient to give a complete rheological characterization; i.e., to predict the relationships between stress and strain for any deformation. For example, if we subject these materials to steady simple shearing and calculate their viscosities using Equation 1-9, we find that the viscosity is not equal to a constant, but varies with shear rate. Thus, this test no longer yields a "material constant," but a "material function":

$$\eta = \eta(\dot{\gamma}).$$

The increased complexity of the structure of these liquids makes it possible for the structure to vary with the shear rate, and this change in structure results in a change in the viscosity. At sufficiently low shear rates, we would expect the variation in structure to become insignificant, so that the dependence of viscosity on shear rate would disappear, unless the material has a yield stress. The value of the viscosity in this range of shear rates is called the "zero shear rate viscosity" and is given the symbol η_0.

If the viscosity decreases as the shear rate is increased, the fluid is said to be "pseudoplastic," and this is the type of behavior most often observed. If the viscosity increases as the shear rate is increased, it is

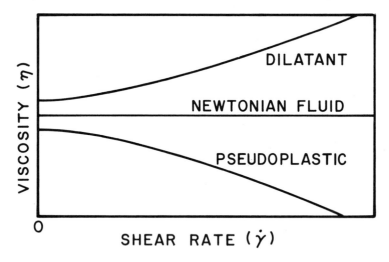

Figure 1-6. Viscosity functions for several types of fluid.

said to be "dilatant." These types of behavior are illustrated in Figure 1-6.

A liquid for which the viscosity function alone provides a complete rheological characterization has been called a "generalized Newtonian fluid."* However, it seems unlikely that any actual fluids exhibit this type of behavior except as an approximation of more complex behavior valid only for certain types of deformation. Liquids which are non-Newtonian generally exhibit not only a shear rate dependent viscosity, but also one or more of the following phenomena.

1. Plasticity.
2. Time dependent structure.
3. Viscoelasticity.

Each of these phenomena will now be illustrated by reference to an experiment involving simple shear.

Plasticity is the phenomenon exhibited when a material is able to withstand a certain amount of stress, its "yield stress," without flowing. Above this yield stress, the material deforms continuously like a fluid. If we subject a plastic material to a simple shearing deformation

---

*Bird *et al.* [1] have discussed the use of this concept in great detail.

employing a range of shear rates, we will find that as the shear rate, $\dot{\gamma}$, approaches zero, the shear stress, τ_{21}, approaches a finite limiting value, which is the yield stress. This type of behavior is shown in Figure 1-7, where it is compared with Newtonian and pseudoplastic types of behavior.

If the adjustment of the structure of the fluid to changes in shear rate requires a significant amount of time, we will find that the viscosity measured in a simple shear experiment varies not only with shear rate, but also with time.

$$\eta = \eta(\dot{\gamma}, t)$$

If the viscosity decreases with time, the material is said to be "thixotropic." If the viscosity increases with time, it is said to be "rheopectic." Thixotropy is the more common type of behavior. These phenomena are compared with Newtonian behavior in Figure 1-8.

In thixotropic and rheopectic materials, the structure (and thus the viscosity) depends on the shearing which has occurred in the past, but

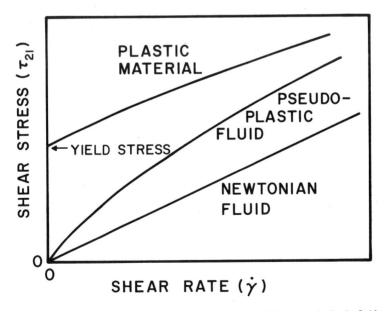

Figure 1-7. Shear stress versus shear rate for a plastic material, a pseudoplastic fluid and a Newtonian fluid.

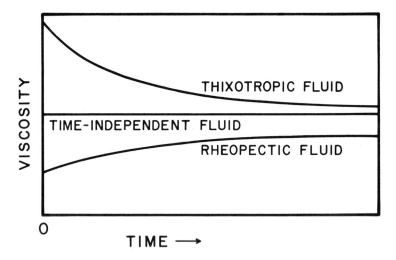

Figure 1-8. Viscosity versus time for materials with time-dependent structure.

in a "viscoelastic" material, the material has a directional "memory" of past deformations. This gives rise to phenomena such as stress relaxation and elastic recoil. Consider, for example, an experiment in which we subject a material to steady simple shearing and then suddenly stop the shearing action and hold the boundary plates in position. If the material involved is a Newtonian fluid or any other inelastic fluid, no stress is necessary to hold the plates in position once the shearing action has stopped. Thus, we might say that after cessation of steady shear, the stress relaxation is instantaneous in inelastic materials. However, if the material studied is viscoelastic, the stress does not return to zero instantaneously, and we obtain a stress relaxation curve of the type shown in Figure 1-9.

Now consider another experiment in which a steady shearing motion like the one illustrated in Figure 1-4 is suddenly interrupted by the removal of the shear stress from the upper (moving) plate. If the material between the plates is a Newtonian fluid or any other inelastic fluid, the upper plate will simply stop when the force on it is removed. However, if the material involved is a viscoelastic liquid, the upper plate will spring back somewhat, but this "elastic recoil" will not be instantaneous because it will be impeded by viscous mechanisms. Figure 1-10 shows the strain versus time curve for a viscoelastic material exhibiting elastic recoil.

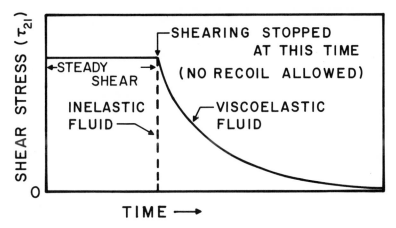

Figure 1-9. Stress relaxation after cessation of steady shear.

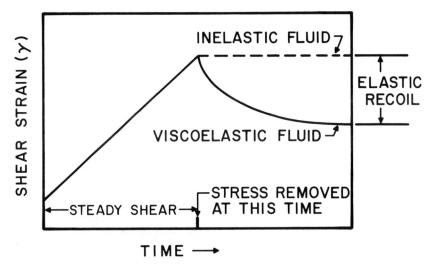

Figure 1-10. Strain recovery (recoil) after elimination of a steady stress.

1.7 WHY POLYMERIC LIQUIDS ARE NON-NEWTONIAN

Polymer solutions and melts are often highly non-Newtonian, and we will examine very briefly now the molecular mechanisms governing the rheological properties of polymeric liquids. While a quantitative exposition of the kinetic theory of macromolecular liquids is not within the scope of this book, it is useful to review some basic qualitative concepts.

First, let us consider, briefly, dilute polymer solutions. If such a liquid is allowed to stand for some time, the polymer molecules will assume an equilibrium distribution of shapes (conformations) and a random distribution of orientations. If the fluid is now subjected to some kind of deformation (for example, if it is sheared or stretched at a significant rate), the polymer molecules will have forces applied to them that will change their shapes so that there will be a shift away from the equilibrium distribution of conformations. An example of what is meant by a change in conformation is shown in Figure 1-11.

In addition, the molecules may align themselves to some preferred flow axis so that the distribution of orientations is no longer random. It is the dependence of conformation and orientation on the flow that gives polymer solutions their complex rheological properties. The change in conformation makes the material properties such as viscosity dependent on the rate of deformation, and since these changes do not occur instantaneously, they make the fluid time dependent. The flow induced orientation of molecules causes the material to exhibit viscoelasticity.

In the case of concentrated solutions and melts of high molecular weight polymers, there is considerable indirect evidence that the rheo-

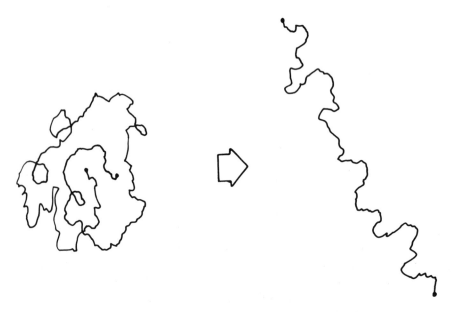

Figure 1-11. Change in conformation of a polymer molecule.

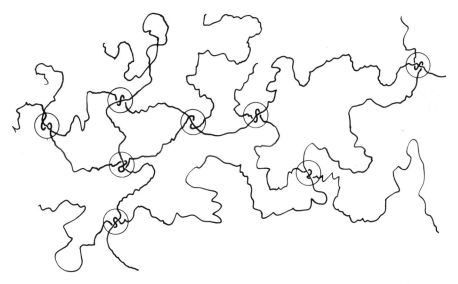

Figure 1-12. Entanglement network as imagined by some rheologists. (Temporary junctions are circled.)

logical behavior is governed by very strong interactions between molecules.* These interactions are so strong that they sometimes result in behavior like that of a crosslinked rubber. However, the points along the molecules where these strong interactions occur seem to be able to move about somewhat in response to deformations. Furthermore, in any interval of time, some of the points of strong interaction seem to disappear while new ones are being formed at other locations. These observations have led to the suggestion that these points of strong interaction can be looked upon as molecular "entanglements," as shown in Figure 1-12, and that these entanglements act as temporary crosslink junctions.

The presence of the temporary network results in rubber-like elastic behavior over short periods of time. However, since the network is not permanent, and old "junctions" are constantly being lost and new ones formed, permanent deformation is possible, and the material's "memory" of past strain fades with time. We would expect the fluid-like aspects of the rheological behavior to be most evident when fairly constant stresses are applied over relatively long periods of time.

---

*This evidence has been summarized by Ferry [2].

We can now interpret the rheological phenomena described in the previous section in terms of the concept of an entanglement network. The decrease in viscosity with shear rate, for example, can be interpreted in terms of an entanglement density which varies with shear rate. Thus, the shearing process is thought to increase the rate of loss of existing entanglements, but not the rate of generation of new ones, so that the number of entanglements in a given volume of material (the entanglement density) has lower equilibrium values at progressively larger shear rates.* Furthermore, since it takes time for the entanglement density to reach an equilibrium value once shearing has been initiated, melts and concentrated polymer solutions exhibit a time dependent structure.

Since the segments of molecules linking the temporary "network junctions" or "entanglements" are flexible, the network itself is elastic, and this gives rise to elastic behavior. A dramatic manifestation of the presence of this network can be observed by subjecting an amorphous polymer to a large stretching strain at a temperature just above the glass transition temperature, where rates of loss and formation of "entanglements" are very small. The deformation is almost completely reversible if the deforming stress is maintained for only a short time before being removed.

The complex rheological behavior of polymeric liquids has two important practical consequences. First, no single rheological property gives a complete rheological characterization of the material, and, second, the measurement of a rheological property requires more careful control of the flow than is necessary to measure the viscosity of a Newtonian fluid.

Thus, in order to select an instrument to study the rheology of a polymeric liquid, one must answer two basic questions:

1. What property should be measured?
2. What deformation (flow) should be used to determine that property?

Having made these basic decisions, there still remains the subsidiary question of whether a suitable instrument is commercially available,

---

*Graessley [3] has made considerable use of this idea in the development of relationships between the viscosity function of a melt and its molecular weight distribution.

and, if not, whether it is feasible to design an instrument expressly for the application at hand. It is hoped that this book will serve as a guide in the answering of these questions.

1.8 SPECIAL PROBLEMS IN THE STUDY OF POLYMER MELTS

The technique used to measure rheological properties depends very much on the general characteristics of the material to be studied. From this point of view, it is possible to classify polymeric materials as follows.*

A. Low viscosity, viscoelastic liquids; e.g., a dilute solution.
B. High viscosity, viscoelastic liquids; e.g., a molten polymer.
C. Soft viscoelastic solids; e.g., a lightly crosslinked polymer (elastomer) or a partially crystalline polymer above its glass transition point.
D. Hard viscoelastic solids; e.g., a glassy polymer or a highly crosslinked system.

There is some overlap between these categories in that some techniques are applicable to materials of two or more types. However, each class of materials has associated with it some unique problems, and we want here to focus our attention on those which arise in the study of high viscosity, viscoelastic liquids (i.e., molten polymers). The unique problems arising in the design of melt rheometers are related to polymer melts being highly non-Newtonian, having very high viscosities and needing to be studied at elevated temperatures.

Because melts are highly non-Newtonian, we must know both the stress and the strain (or strain rate) in the material in order to interpret experimental data in terms of well-defined material properties. Thus, the simple viscometers useful for Newtonian fluids cannot be used to study melts. In addition, the high stresses necessary to deform polymer melts require that the apparatus used be very robust and stiff. Any bending or twisting of the apparatus can result in a change in the dimensions of the sample under study, and this will affect the accuracy of the measurement.

---

*This method of classification is based on that used by Ferry [2].

Bulk polymers of commercial significance are usually solid-like at room temperature and are processed in the melt state at temperatures in the range of 150°C to 250°C. These circumstances pose problems for the rheometer designer, as loading the sample and removing it after testing are normally carried out at temperatures below the melt temperature. As a result, specially molded preforms are usually used as samples, and special provision must be made for removing polymer after a test.

The high temperatures required also pose problems in temperature control. During the melting stage, a high rate of heating is desired, but once the operating temperature is reached, precise control of temperature is required. The high viscosity of the melt implies a high rate of viscous dissipation unless strain rates are very low. This will result in a significant nonuniformity of temperature within the sample unless the sample thickness is very small.

Finally, the use of elevated temperatures leads inevitably to chemical degradation of the sample, and this means that the useful lifetime of a given sample in the rheometer is generally rather short.

1.9 OTHER BOOKS OF INTEREST TO THE MELT RHEOLOGIST

The scope of this book is limited to experimental techniques and apparatus useful in the study of molten polymers. However, in the planning of an experimental program or in the interpretation and use of rheological data, the reader may wish to refer to recent books on closely related topics. Some of the most useful of these are listed below.

- Polymer rheology
    Ferry [2]
    Vinogradov and Malkin [4]
    Bird, Hassager, Armstrong and Curtis [5]
- Polymer melt rheology and/or plastics processing
    Han [6]
    VDI (in German) [7]
    Tadmor and Gogos [8]
    Throne [9]
    Middleman [10]
    Pearson [11]

- Experimental methods
  Whorlow [12]
  Macosko and Tirrell [13]
  Walters [14]
  Walters [15]
  Leblanc (in French) [16]
- Formulation and use of constitutive equations
  Bird, Armstrong and Hassager [1]
  Schowalter [17]
  Astarita and Marrucci [18]
  Lodge [19, 20]
- Extensional flows
  Petrie [21]

Other important sources of information for rheologists are the two periodical bibliographies noted below.

- *Rheology Abstracts*, published quarterly by the British Society of Rheology and distributed to all members.
- *Dokumentation Rheologie*, published annually by the Bundesanstalt für Materialprüfung Fachgruppe Rheologie und Tribologie and the DRG, Unter den Eichen 87, D-1000 Berlin 45.

Research periodicals containing a significant number of papers on melt rheology include the following.

- Published in the United States:
  - *Journal of Rheology*,* published by John Wiley & Sons, New York, for the Society of Rheology.
  - *Polymer Engineering and Science*, published by the Society of Plastics Engineers.
  - *Journal of Applied Polymer Science*, published by John Wiley & Sons, New York.
  - *Journal of Polymer Science*, Polymer Physics Edition, published by John Wiley & Sons, New York.

---

*Formerly called *Transactions of the Societ of Rheology*.

- Published in Europe:
  - *Journal of Non-Newtonian Fluid Mechanics*, edited in the U.K., published by Elsevier, Amsterdam.
  - *Rheologica Acta*, published by Dr. Dietrich Steinkopff Verlag, Darmstadt, Federal Republic of Germany.
  - *Kunststoffe*, published by Carl Hauser Verlag, Munich, Federal Republic of Germany (special edition, subtitled "German Plastics," contains English translations of technical articles).

Chapter 2
Measurable Rheological Properties

This book deals primarily with the measurement of well-defined physical properties of highly viscous, viscoelastic liquids. These properties are defined in terms of the response of the liquid to a known deformation. In order for the deformation to be known, in spite of the variation of rheological properties from one liquid to another, the flow used must be "controllable." This means it must be a flow in which, by prescribing certain boundary tractions or motions, a certain type of deformation will always occur, no matter what liquid is involved.

In addition, since molten polymers are viscoelastic, stresses on a fluid element at time t depend not only on the rate of deformation of that element at time t, as in the case of a Newtonian fluid, but also on the recent past rate and pattern of deformation. Thus, we need to control not only the deformation at one point in the rheometer, but the deformation history of a fluid element over some finite length of time.

It might at first be thought that any "steady" flow would be useful as the basis of a rheological technique. However, a flow which is "steady," in the sense that at each point in the flow the velocity is independent of time, does not necessarily provide a controlled strain history. For example, consider the flow at the entrance to a capillary (as shown in Figure 2-1). First we note that the strain rate and the pattern of the deformation experienced by a fluid element change markedly as that element moves along the streamlines. Furthermore, these will vary from one material to another, so that changing the material also changes the deformation history in an uncontrollable way. For these reasons, rheometrical flows are most often uniform flows, and only properties defined in terms of uniform flows will be presented in this chapter.

It should be pointed out here that the actual flows used in the laboratory to measure these properties never reproduce exactly the ideal-

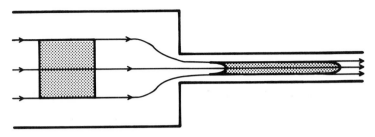

Figure 2-1. Deformation of a fluid element at the entrance to a capillary.

ized, uniform flows on the basis of which the properties are defined. Thus, there are always sources of error, such as end and edge effects, which must be taken into account in the design and interpretation of experiments. Methods for minimizing the effects of such deviations from uniform flow are described in Chapters 4 and 5.

The simplest example of a uniform, controllable flow is simple shear, a flow defined in Chapter 1. If the gap is small so that inertia can be neglected, or if we operate at steady state, the motion of the upper plate in Figure 1-4 with a velocity V_w will always produce a uniform shearing deformation in the liquid, with the shear rate given by Equation 1-6. Measurement of the force exerted on the plate and its area allows us to calculate the viscosity, and if we repeat the test using a variety of plate speeds, we can determine the viscosity function, $\eta(\dot{\gamma})$.

2.2 PROPERTY MEASUREMENT VERSUS EMPIRICAL TESTING

While flow between parallel plates of infinite extent is not a practical basis for a laboratory test, we will see in Chapters 4 and 5 that there are several, practically realizable shear flows that can be used to determine the viscosity. No matter which of these flows is used, the same value of viscosity will be found at a given shear rate. Thus, the significance of a curve of viscosity versus shear rate is based on the basic definition of the property and is independent of the details of the experimental method.

In contrast to flows in which the deformation is controlled and a well-defined material property is determined, a number of empirical tests are used for comparing materials, in which one has little control or knowledge of the flow. Indeed, the flow pattern may vary from one material to another. Thus, the strain history is different for each material studied, and it is, in a sense, a different property that is measured in each case.

For example, consider a test which we might devise to determine a characteristic of a melt we will call, for purposes of this example, its "fluidity index." As shown in Figure 2-2, a cylindrical tank containing some of the melt to be studied is fitted with a piston of sufficient weight to force the melt through an orifice in the bottom of the tank at a noticeable rate. The amount of material flowing through the orifice in a certain length of time is measured, and this is our measure of the fluidity index.

First we note that we cannot control the details of the flow pattern and that this flow pattern depends strongly on the rheological properties of the fluid. Thus, the strain history varies from one fluid to another, and our fluidity index is not a well-defined physical property. Furthermore, if this technique is to be used in different laboratories, and we wish the results to be comparable, we must take great care that all the details of the apparatus and test procedure are exactly the same in each laboratory. Thus, the fluidity index is not defined in terms of a particular deformation, but in terms of a particular, arbitrary laboratory procedure.

Finally, we observe that melts which have quite distinct rheological properties can yield equal values of the fluidity index. Thus, since different combinations of properties can give the same value of fluidity

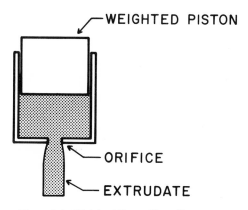

Figure 2-2. Weight-driven orifice efflux tester.

index, the use of this parameter as a measure of the similarity of materials can be highly misleading.*

Sometimes, a laboratory test is carried out, not because it leads to the determination of a particular physical property, but because it simulates some process of commercial importance. For example, wind tunnel tests allow an aerodynamicist to determine the drag that will be experienced by an airplane by measuring the drag force on a scale model of the plane. To make quantitative use of data obtained from such a simulation experiment, we must know the scaling laws relating the laboratory simulation to the full-scale process. In the case of aerodynamics, we make use of dimensionless groups such as Mach number, Reynolds number and drag coefficient to plan the simulation and to interpret the results of the experiment.

The success of the simulation procedure depends on our ability to specify all the relevant fluid properties in terms of a few material constants such as viscosity and density. However, for complex flows of molten polymers, the relevant rheological properties cannot be specified in terms of a few material constants so that simple scaling laws cannot be established. While it may be found that a particular laboratory test is useful for predicting the behavior of a melt in some complex flow of industrial importance, the usefulness of the test must be demonstrated in the field and cannot be established by *a priori* dimensional analysis. This means that the use of such an empirical simulation, for a class of polymers other than that for which the validity of the simulation has been demonstrated, is a highly unreliable procedure. An example of an empirical simulation test is the "melt strength" test described in Chapter 7.

As has been pointed out, the measurement of well-defined material properties provides a more reliable basis than an empirical test for comparing materials or for analyzing a process by use of fundamental physical principles. It needs to be emphasized that for viscoelastic materials such as molten polymers, no single rheological property gives a complete picture of the material behavior. Thus, if one wishes to evaluate resins for use in a particular process, the decision as to which rheological properties to measure must be based on the types of flow thought to be of

---

*The particular test described here is similar in many ways to the "melt index" test often used to evaluate molten polymers. However, a melt indexer has a tube in place of the orifice, and this makes it possible to use this instrument for crude measurements of viscosity if a sufficiently long tube is used. The melt index test is described in detail in Chapter 4.

central importance in the process. This question of selecting the properties to be measured in a particular application is taken up in Chapter 8.

Of course, if we knew with certainty that a particular molten polymer behaved according to a particular constitutive equation containing a few material constants, then once we had done sufficient experiments to determine those material constants, we could predict with confidence how the material would behave in any type of situation. Unfortunately, although considerable effort has been directed at the problem, no constitutive equation has been devised which is generally valid for all flows of all molten polymers. Thus, each time we measure an additional material function for a melt, we obtain new information as to its rheological behavior, and there is no way that we can predict with confidence exactly how the melt will behave in a flow other than the ones already studied.

In the remainder of this chapter, a number of rheological properties* will be defined. The practical problems associated with the measurement of these properties will be discussed in succeeding chapters.

2.3 TIME-INDEPENDENT MATERIAL FUNCTIONS

A large number of material functions may be defined, and it will be useful to have some criteria that can be used to decide which ones are likely to be useful. First, of course, it must be possible to determine the function in a practically realizable experiment. Secondly, it is desirable that the number of independent variables be limited to one, or at most two. For example, the viscosity is a function of only one variable, the shear rate, and it can, therefore, be represented by a single curve. The more parameters there are associated with the function, the more difficult it gets to represent the function in a way that facilitates the comparison of materials by visual inspection of graphs.

Since molten polymers are viscoelastic, they have a "memory"; i.e., their behavior at any instant depends on how they have been deformed in the past. This means that, in general, time must be included as an independent variable for a rheological test and, thus, a parameter of the material function determined by means of that test. However, since melts are basically liquids rather than solids, their memory "fades" with

---

*The term "rheological property" is used in this book to mean a measurable material function.

time, and the current behavior of the material is practically independent of deformations which occurred sufficiently far in the past.

For this reason, there are certain deformations for which material functions can be measured that do not have time as an independent variable, and these are called "motions with constant stretch history."* In this type of flow, the deformations of a given fluid element that occur in successive equal intervals of time are identical. Steady simple shear is an example of a motion with constant stretch history. Since viscoelastic liquids have a fading memory, only deformations in the relatively recent past affect the present behavior of the material, and in a motion with constant stretch history, that portion of the strain history affecting the present behavior remains the same as time passes. Thus, the rheological behavior becomes independent of time, and the only parameter of material functions measured by the use of these flows is a strain rate. Two types of motion with constant stretch history can be generated in the laboratory and have been used extensively for rheological studies. These are viscometric flow and steady extension, and the material functions that can be measured by use of these types of flow are defined in the following sections.

2.3.1 Viscometric Functions

A "viscometric flow" is one in which the deformation experienced by a given element of fluid is indistinguishable from steady simple shear.** Obviously, steady simple shear is itself a viscometric flow, but the use of two plates of infinite extent is clearly not a practical basis for a laboratory test. Viscometric flows that are useful in the laboratory include the flow in a tube (away from the ends) and the flow between concentric cylinders, one of which is rotating. Flows that are approximately viscometric include the flow between a stationary disk and a rotating cone or disk. These flows will be discussed in detail in Chapter 5, and we wish at this point simply to define the material functions which can, in principle, be determined by use of a viscometric flow.

The definition of the viscosity was given by Equation 1-9, but we will

---

*Lodge and Walters [22] have reviewed the origin and use of this concept.
**Coleman, Markovitz and Noll [23] have written a comprehensive treatment of the mathematical basis and experimental use of viscometric flows.

rewrite it here using the traditional nomenclature in which the shear stress, τ_{21}, in simple shear, is written simply as τ.

$$\eta \equiv \tau/\dot{\gamma} \tag{2-1}$$

For a Newtonian fluid, this property is independent of shear rate, but for polymeric liquids, the viscosity depends on the shear rate and is thus a material function.

$$\eta = \eta(\dot{\gamma})$$

We now wish to ask if there are any other material functions that can be determined in a simple shear flow. So far, we have made use only of the shear stress, but we might consider what information could be obtained by measuring the normal stress components. For the case of a Newtonian fluid, we can use Equation 1-11 to show that the normal components of the extra stress are all equal to zero.

$$\tau_{11} = \tau_{22} = \tau_{33} = 0$$

However, in polymeric liquids, the shearing motion has the effect of producing some orientation of the molecules so that the liquid properties become anisotropic; i.e., dependent on direction. The deviation of the distribution of molecular orientations from the random distribution, which is attained when the fluid is at rest, represents a departure of the system from its unstressed state, and in order to maintain this departure, it is necessary that the stress be anisotropic. In other words, the anisotropy of the fluid structure resulting from the shearing motion gives rise to normal stress differences.

$$\tau_{11} \neq \tau_{22} \neq \tau_{33}$$

It should be recalled from Chapter 1 that these normal components of stress have no absolute rheological significance. Thus, we can define only two additional material functions, and these are the "first and second normal stress differences" usually defined as follows.*

---

*In some books and articles, alternate definitions are given as follows: $N_1 = \tau_{11} - \tau_{22}$; $N_2 = \tau_{11} - \tau_{33}$.

$$N_1 \equiv \tau_{11} - \tau_{22} \qquad (2\text{-}2)$$
$$N_2 \equiv \tau_{22} - \tau_{33} \qquad (2\text{-}3)$$

Like the viscosity, these rheological properties are functions of the shear rate. Thus, we conclude that three independent material functions can be determined in a viscometric flow. These "viscometric functions" are

$$\eta(\dot{\gamma}), \ N_1(\dot{\gamma}), \ N_2(\dot{\gamma}).$$

If a material is intrinsically isotropic, as is the case with most molten polymers, a reversal in the direction of flow in simple shear should result in steady state stresses having the same magnitude as before, and the viscometric functions will be unchanged. This symmetry can be expressed as follows.

$$\eta(\dot{\gamma}) = \eta(-\dot{\gamma}) \qquad (2\text{-}4a)$$
$$N_1(\dot{\gamma}) = N_1(-\dot{\gamma}) \qquad (2\text{-}4b)$$
$$N_2(\dot{\gamma}) = N_2(-\dot{\gamma}) \qquad (2\text{-}4c)$$

It is of interest to see how the viscometric functions behave at very low shear rates. First, we observe that at zero shear rate, the shear stress and the normal stress differences are equal to zero:

$$\lim_{\dot{\gamma}\to 0} \tau = \lim_{\dot{\gamma}\to 0} N_1 = \lim_{\dot{\gamma}\to 0} N_2 = 0. \qquad (2\text{-}5)$$

Moreover, in Chapter 1, it was argued that at very low, but finite, shear rates, the structure of polymeric liquids might be expected to be independent of shear rate. This behavior is consistent with the concept of fading memory, because if the shear rate is sufficiently small, the total strain which has occurred recently enough to have an effect on the current behavior of the fluid is negligible. This line of reasoning leads to the following conclusion regarding the limiting behavior of the viscosity as the shear rate approaches zero.

$$\lim_{\dot{\gamma}\to 0} \left(\frac{\partial \eta}{\partial \dot{\gamma}} \right) = 0 \qquad (2\text{-}6)$$

This implies that the viscosity has a limiting value at zero shear rate, and this value is called the "zero shear viscosity."

$$\lim_{\dot{\gamma}\to 0} \eta = \eta_0 \tag{2-7}$$

If we make use only of qualitative ideas of melt structure and fading memory, we cannot say any more about the behavior of the viscometric functions. However, there is a line of reasoning in the science of rational mechanics which allows us to go one step further. Rational mechanics is a branch of applied mathematics in which one starts from very general hypotheses concerning the mechanical behavior of a certain class of materials and examines the implications of these hypotheses in terms of the behavior of the rheological material functions of such materials. In the case of viscoelastic liquids, this procedure leads to the conclusion that at very small, finite shear rates, the viscosity should be independent of shear rate, and the normal stress differences should be proportional to the square of the shear rate.*

The conclusion regarding the viscosity is equivalent to the behavior expressed by Equation 2-6, but the conclusion regarding the normal stress differences, while consistent with Equation 2-5, goes beyond it. This latter result has inspired the definition of two "normal stress coefficients" which can be used in place of the normal stress differences. These are defined as follows.

$$\Psi_1 \equiv N_1/\dot{\gamma}^2 \tag{2-8a}$$
$$\Psi_2 \equiv N_2/\dot{\gamma}^2 \tag{2-8b}$$

The advantage of the use of these material functions, in place of N_1 and N_2, is that they may be expected to have finite limits as the shear rate approaches zero.

$$\lim_{\dot{\gamma}\to 0} (\Psi_1) \equiv \Psi_{1,0} \tag{2-9a}$$
$$\lim_{\dot{\gamma}\to 0} (\Psi_2) \equiv \Psi_{2,0} \tag{2-9b}$$

---

*This result comes from the theory of the "second order fluid" and is discussed in [1, 14, 17, 18].

A word of caution is in order here regarding the experimental deter-
mination of η_0, $\Psi_{1,0}$ and $\Psi_{2,0}$. Unless the liquid of interest has been studied at sufficiently low shear rates that τ_{21} becomes proportional to $\dot{\gamma}$,
and N_1 and N_2 become proportional to $\dot{\gamma}^2$, rational mechanics provides
no clue as to how to extrapolate the data to obtain the limiting values,
η_0, $\Psi_{1,0}$ and $\Psi_{2,0}$. Thus, in the absence of a constitutive relationship
believed to be valid for the liquid in question, such extrapolations are
arbitrary and cannot yield meaningful values of η_0, $\Psi_{1,0}$ and $\Psi_{2,0}$. This
point is especially relevant in the case of molten polymers, because it
often proves practically impossible to operate rheometers at shear rates
sufficiently low so that η, Ψ_1 and Ψ_2 reach their limiting values.

It is often convenient to be able to express the viscometric functions
in terms of explicit equations. Over certain intermediate ranges of shear
rates, the following "power law" models are often found to give a reasonably good representation of experimental data.*

$$\eta = k\dot{\gamma}^{n-1} \tag{2-10}$$
$$N_1 = c\dot{\gamma}^{m} \tag{2-11}$$

These particular forms, however, are inconsistent with some of the basic
properties of the viscometric functions which were presented earlier in
this section. For example, Equation 2-10 does not have the low shear
rate limiting behavior defined by Equation 2-6, and neither power law
expression has the symmetry property defined by Equation 2-4 unless
the quantities $(n - 1)$ and m are even integers. Thus, these are simply
empirical approximations valid only at relatively large shear rates. In
an effort to extend the range of these power laws, many more complex
forms have been proposed, especially for the viscosity function. A number of these have been described and compared by Elbirli and Shaw
[24] and by Bird et al. [1].

2.3.2 Extensional Flows

Another type of deformation that can be used to obtain additional information concerning the rheological behavior of a viscoelastic liquid is

---

*The second normal stress difference for melts is usually found to be much smaller than the first,
and its precise measurement is rather difficult. Thus, very few sets of reliable data have been
published.

extensional flow. Consider, for example, "simple extension," also called "uniaxial extension." This type of deformation is illustrated in Figure 2-3. It might be generated by preparing a rod-shaped sample, fixing one of its ends, and applying a tensile force to the opposite end. The challenging problems associated with performing such a test on a liquid-like material will be described in Chapter 6.

Quantities which can be measured in simple extension include the stretching force, F, and the length, L, both of which may be functions of time. If the material is incompressible, the volume will remain constant so that the product of the length and area will remain equal to its initial value.

$$LA = L_0 A_0$$

Now we wish to inquire how these data can be interpreted so as to yield well-defined material properties and, in addition, how to carry out the test so that a motion with constant stretch history is generated.

Considering first the force, it is obvious that this should be divided by the cross-sectional area to yield a stress. Since the applied force is a tensile force, this stress will be a positive normal stress. As was explained in Chapter 1, it is not the normal stresses themselves that are meaningful as rheological variables, but differences in the normal stresses. In the present case, let x_1 be the direction of stretching, and let x_2 be any direction normal to this. Since this deformation has axial symmetry, the particular choice of direction for the x_2 axis is not important, and $\tau_{22} = \tau_{33}$. Thus, a rheologically meaningful variable is the normal stress difference, $\tau_{11} - \tau_{22}$.

Turning now to the question of how to define a meaningful measure of strain, it is clear that the amount of elongation resulting from a given

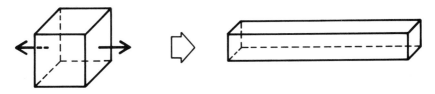

Figure 2-3. Simple (uniaxial) extension.

stress will depend on the length of the sample, and we might think of defining the infinitesimal extensional strain as

$$\frac{dL}{L_0}$$

where L_0 is the initial length of the sample. However, for materials that are liquid-like (i.e., those with fading memory), the relevance of the initial length for the current material behavior continually decreases as the stretching continues. For these materials, the only reference length that continues to have the same importance from one instant to another is the current length, L.

Thus, we can define a measure of strain that is meaningful for liquids as follows.

$$d\epsilon = dL/L$$

By integration, the strain for finite deformation can be shown to be

$$\epsilon = \ln(L_2/L_1). \tag{2-12}$$

By comparison with Equation 1-1 we see that this is the Hencky strain defined in Chapter 1. The strain rate is given by

$$\dot{\epsilon} \equiv \frac{d\epsilon}{dt} = \frac{1}{L}\frac{dL}{dt} = \frac{d\ln L}{dt}. \tag{2-13}$$

To relate the Hencky strain rate to the experimental variable L, we note that dL/dt is the velocity, V, at the end of the sample, so that Equation 2-13 can be rewritten as follows.

$$\dot{\epsilon} = V/L \tag{2-14}$$

If the strain rate, $\dot{\epsilon}$, is maintained constant, we obtain a motion with constant stretch history, and this deformation is called "steady simple extension" or "steady uniaxial extension."

It is of interest to derive equations for the velocity distribution for simple extension, as this will lead to a general definition of an extensional flow. To this end, note that since the deformation is homogeneous, the strain rate will be the same for all elements of the specimen. For example, consider an element of material extending from the fixed end, where $x_1 = 0$, to some point x_1, less than L. This element should experience the same strain rate as the sample as a whole. Thus, if v_1 is the velocity at x_1, we have

$$v_1 = \dot{\epsilon} x_1.$$

The velocity components v_2 and v_3 can be shown to be proportional to x_2 and x_3, respectively, and this is the general characteristic of all extensional flows, which can be defined as flows having velocity distributions of the following type.

$$v_1 = a_1 x_1 \qquad (2\text{-}15\text{a})$$
$$v_2 = a_2 x_2 \qquad (2\text{-}15\text{b})$$
$$v_3 = a_3 x_3 \qquad (2\text{-}15\text{c})$$

The a_i parameters can be functions of time, but the law of conservation of mass for an incompressible material can be used to show that they must sum to zero.

$$a_1 + a_2 + a_3 = 0 \qquad (2\text{-}16)$$

This leaves us free to choose arbitrary values for any two of the three parameters.

If we make these parameters independent of time, the resulting deformation will be a motion with constant stretch history. There are an infinite number of such flows, but only three of these have been used as bases for rheological tests, and these will be described in the next three sections.

2.3.3 Simple Extension and the Extensional Viscosity

Noting that simple (uniaxial) extension is axially symmetric, we can see from Equations 2-15 and 2-16 that this flow has the following velocity distribution.

$$v_1 = \dot{\epsilon} x_1 \qquad \text{(2-17a)}$$
$$v_2 = - \tfrac{1}{2} \dot{\epsilon} x_2 \qquad \text{(2-17b)}$$
$$v_3 = - \tfrac{1}{2} \dot{\epsilon} x_3 \qquad \text{(2-17c)}$$

An alternative representation can be written by making use of cylindrical coordinates.

$$v_z = \dot{\epsilon} z \qquad \text{(2-18a)}$$
$$v_r = - \tfrac{1}{2} \dot{\epsilon} r \qquad \text{(2-18b)}$$
$$v_\theta = 0 \qquad \text{(2-18c)}$$

If $\dot{\epsilon}$ is independent of time, the flow is steady simple extension, and since this is a motion with constant stretch history, it should have associated with it a material function in which times does not appear as an independent variable. Since the flow is axially symmetric, the stress is also symmetric with

$$\tau_{22} = \tau_{33} = \tau_{rr}.$$

Thus, there is one independent normal stress difference, and a material function having the units of viscosity can be defined as

$$\eta_T(\dot{\epsilon}) \equiv \frac{\tau_{11} - \tau_{22}}{\dot{\epsilon}} = \frac{\tau_{zz} - \tau_{rr}}{\dot{\epsilon}}. \qquad \text{(2-19)}$$

This property is called the "extensional viscosity."*
    For the special case of a Newtonian fluid, Equation 1-11 can be used to show that

$$\eta_T = 3\eta. \qquad \text{(2-20)}$$

Thus, as expected, once the viscosity of a Newtonian Fluid is known, we learn nothing new by subjecting it to an extensional flow.
    For a polymeric liquid, however, we are able to use extensional flows to study aspects of rheological behavior that are not apparent in viscom-

---

*The subscript, T, is in remembrance of F. T. Trouton, who recognized the importance of uniaxial extension early in this century [21].

etric flows because of an important difference between the two types of flow. We note that the rate of separation of two material points in a fluid undergoing steady simple extension can be obtained by integrating Equation 2-13 and interpreting L as the distance between any two material points, measured in the direction of stretching.

$$L(t) = L_0 e^{\dot{\varepsilon}t} \tag{2-21}$$

Thus, the separation of the material points is exponential in time. In steady simple shear, on the other hand, the separation of material points located at different values of x_2 (see Figure 1-4) is proportional to the time at large times. As a result, the tendency of the flow to produce orientation of the polymer molecules is much greater for extensional than for viscometric flows.*

If the extensional strain rate is sufficiently small, we would expect a polymeric fluid to behave like a Newtonian fluid so that

$$\lim_{\dot{\varepsilon} \to 0} [\eta_T(\dot{\varepsilon})] = 3\eta_0. \tag{2-22}$$

In fact, for melts, it is sometimes difficult to carry out an experiment at strain rates sufficiently low for this behavior to be observed. On the other hand, there may arise situations in which it is possible to determine the limiting Newtonian value for η_T when it has proven practically impossible to determine η_0 by use of a shearing flow. In such a case, Equation 2-22 can be used to calculate a reliable value for η_0.

One often finds in the technical literature that when both $\eta(\dot{\gamma})$ and $\eta_T(\dot{\varepsilon})$ have been determined for a given material, both results are plotted on a single graph. This is a way to economize on graph paper, but the comparison of the two functions in this way has no known significance. As was indicated above, the significance of the two strain rates $\dot{\varepsilon}$ and $\dot{\gamma}$ in terms of stretch rates is quite different.

White [26] and Petrie [21] have reviewed the results to date of measurements of $\eta_T(\dot{\varepsilon})$ for molten polymers, and White [26] has related the

---

*Tanner [25] has discussed the importance of the difference between steady shear and steady extension, and has suggested a classification of flows as "weak" or "strong," depending on the rate of separation of material points, with a "strong flow" being the one most likely to produce orientation.

behavior of this function to melt behavior in the melt spinning process and at the entrance to a capillary.

2.3.4 Biaxial Extension

If we could subject a material to the extensional flow described by Equation 2-17, but with the strain rate, $\dot{\epsilon}$, equal to a negative number, we would produce a compressional flow that would deform a sample or an element of material as shown in Figure 2-4. In fact, it is customary to look upon this deformation as radial extension rather than uniaxial compression.

Taking this point of view, we would sketch the deformation as shown in Figure 2-5 and we would define the principal strain rate as follows.

$$\dot{\epsilon}_b \equiv v_r/r \tag{2-23}$$

For a sample in the shape of a circular disk having a radius R, this "biaxial strain rate" would be

$$\dot{\epsilon}_b = \frac{1}{R}\frac{dR}{dt} = \frac{d\ln R}{dt}. \tag{2-24}$$

Furthermore, when this flow is looked upon as biaxial stretching, it is thought of as being generated by a radial, tensile stress rather than an axial, compressive stress. Thus, the "biaxial extensional viscosity" is defined as follows.

$$\eta_b \equiv \frac{\tau_{rr} - \tau_{zz}}{\dot{\epsilon}_b} \tag{2-25}$$

The fact that biaxial extension is equivalent to uniaxial compression might be thought to imply that the two material functions, $\eta_T(\dot{\epsilon})$ and $\eta_b(\dot{\epsilon}_b)$, contain the same rheological information, but this is not true. Consideration of the type of orientation produced in the two cases helps us to understand why different aspects of the rheological properties of the melt are manifested in the two flows. In uniaxial extension, there is a tendency for alignment in a direction parallel to the axis of symmetry, whereas in biaxial extension, the deformation promotes orientation in the plane perpendicular to the axis of symmetry.

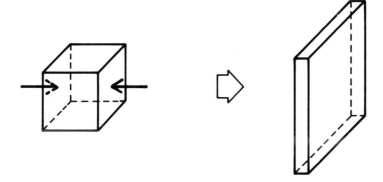

Figure 2-4. Uniaxial compression.

Figure 2-5. Biaxial extension.

For a Newtonian fluid, it can be easily shown that

$$\eta_b = 6\eta. \tag{2-26}$$

Furthermore, the discussion at the end of the previous section on extensional viscosity is equally valid in the present context, and we can conclude that for polymeric liquids,

$$\lim_{\dot{\epsilon}_b \to 0} (\eta_b) = 6\eta_0. \tag{2-27}$$

2.3.5 Planar Extension

One additional extensional flow has been used to study molten polymers, and we will call this "planar extension."* This type of deforma-

---

*This type of deformation has also been called "pure shear," because the change in shape of a fluid element is similar to that which takes place in simple shear. However, in shear there is also a net rotation of the fluid element. Moreover, the *rate* of stretching of a fluid element in the two flows is quite different in its dependence on time.

tion is illustrated in Figure 2-6. As in the case of uniaxial extension, there is a stretching motion in one direction (let us say the x_1 direction), but motion in one of the other two directions (say the x_3 direction) is prevented. To accomplish this, it is necessary to apply some tension, σ_{33}, in the x_3 direction, as indicated by the smaller lateral arrows in Figure 2-6. If we again identify $\dot{\epsilon}$ as the strain rate in the direction of stretching, the velocity components are as follows.

$$v_1 = \dot{\epsilon} x_1 \tag{2-28a}$$
$$v_2 = -\dot{\epsilon} x_2 \tag{2-28b}$$
$$v_3 = 0 \tag{2-28c}$$

Since there is no symmetry, the three normal stress components will take on different values, and there will be two independent rheologically significant normal stress differences. However, only one of these is conveniently measured experimentally, and that is $(\tau_{11} - \tau_{22})$. Thus, we define a "planar extensional viscosity" as follows.

$$\eta_p \equiv \frac{\tau_{11} - \tau_{22}}{\dot{\epsilon}} \tag{2-29}$$

For a Newtonian fluid, we can use Equations 1-11 and 2-28 to show that

$$\eta_p = 4\eta. \tag{2-30}$$

Thus, for a polymeric liquid, we expect that

$$\lim_{\dot{\epsilon} \to 0} (\eta_p) = 4\eta_0. \tag{2-31}$$

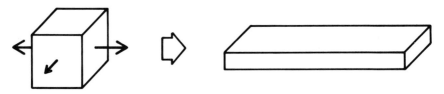

Figure 2-6. Planar extension.

2.4 TIME DEPENDENT MATERIAL FUNCTIONS

Since viscometric flow and steady extensional flows are motions with constant stretch history, the material functions we measure by use of these flows are independent of time. These material functions generally have a strong dependence on strain rate for molten polymers, and this dependence tells us something about how deformation of a constant rate alters the structure of the melt. However, these material functions supply no direct evidence of the melt's elasticity or of the relative effects on the present behavior of strains that took place at various times in the past. To obtain information about the elasticity and strain history dependence of the melt, we must turn to deformations in which the strain rate varies with time.

There are any number of deformations we might choose as the basis of a rheological test, and we usually pick ones that can be described in terms of only one or two parameters. For example, in a traditional stress relaxation experiment, we take a sample of material that has not been deformed for a long time and subject it to a sudden strain, γ_0, in the shortest interval of time practicable, and then observe the shear stress as a function of time. This is called a "stress relaxation" experiment. The strain pattern and typical stress relaxation curves for an elastomer and a melt are compared in Figure 2-7. The shear stress in the elastomer has a non-zero, steady state value, but it falls eventually to zero in the melt because the melt has a fading memory of the past strain due to the impermanence of its entanglement network.

Once we have specified that we are doing a stress relaxation test in shear, we need only specify one parameter, γ_0, to describe exactly the deformation pattern. Thus, the shear stress is a function of time and strain amplitude.

$$\tau = f(t, \gamma_0)$$

Following the terminology used to described elastic solids, it is customary to represent the stress relaxation material function in terms of a relaxation modulus defined as follows.

$$G \equiv \tau/\gamma_0 \qquad (2\text{-}32)$$

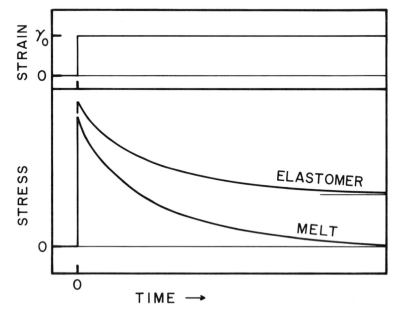

Figure 2-7. Stress relaxation after sudden straining for a crosslinked elastomer and for a melt.

Thus, the results of a stress relaxation experiment are usually given in the form of a plot of G versus t with γ_0 as a parametric variable:

$$G = G(t, \gamma_0).$$

In fact, this type of stress relaxation experiment is more often employed to study solid-like materials than liquids, and deformations more useful for polymer melts will be described in more detail in a later section. Meanwhile, the relaxation experiment will provide a useful basis for a brief discussion of linear viscoelasticity.

2.4.1 Linear Viscoelasticity

If the strain, γ_0, in the experiment described above is sufficiently small, the relaxation modulus is independent of γ_0.

$$\tau(t)/\gamma_0 = G(t) \tag{2-33}$$

This implies that the stress, measured at any particular time t_1, in a series of experiments in which the strain is varied (but kept small), is proportional to the strain.

$$\tau(t_1) = G(t_1)\gamma_0 \qquad (2\text{-}34)$$

This kind of behavior is called "linear viscoelasticity."

Linear viscoelastic behavior, in which a material function becomes independent of the amplitude of the strain or the strain rate used in an experiment, is often observed when the strain or strain rate is sufficiently small. This is not unexpected in the light of our understanding of the structure of polymeric liquids. Nonlinear behavior results from the fact that the deformation has altered the state of the polymer molecules. However, if the strain is sufficiently small, the molecular structure will be practically unaffected, and linear behavior will be observed. In experiments in which a continuous, but time varying, shearing deformation is employed (rather than the one-time strain used in the stress relaxation experiment), we will observe linear viscoelasticity as long as the strain rate is very small. This can be explained in terms of the concept of fading memory. If the motion is always sufficiently slow, the total strain that has occurred during the length of time over which past deformations can affect the present behavior will be too small to alter the molecular structure.

Clearly, a study of the linear viscoelastic properties of a material tells us only about its behavior in the unstrained state in which the distributions of molecular conformation and orientation and the entanglement density have their equilibrium values. The advantage of the use of linear viscoelasticity to characterize a material is that the representation of the material functions is simplified by the disappearance of the strain or strain rate parameter. However, we must keep in mind that these properties give no direct evidence of the ways in which large deformations can alter the structure and, thus, the rheological behavior of polymeric materials.

The various material functions that can be defined for materials exhibiting linear viscoelasticity are all, in principle, related to one another. Various exact and approximate forms of these relationships have been enumerated by Ferry [2].

2.4.2. Stress Growth and Relaxation in Simple Shear

A strain history that is fairly easy to generate in the laboratory and to describe by use of a single parameter is one in which, starting with a sample that has been underformed for a long time, the material is suddenly subjected to shearing at a steady rate, $\dot{\gamma}_0$, and the shearing continued until the shear stress and normal stress differences reach their steady state values. This is called a "stress growth experiment," and the resulting data are represented in terms of "stress growth functions" defined as follows.

$$\eta^+(t, \dot{\gamma}_0) \equiv \tau/\dot{\gamma}_0 \tag{2-35}$$
$$N_1^+(t, \dot{\gamma}_0) \equiv \tau_{11} - \tau_{22} \tag{2-36}$$
$$N_2^+(t, \dot{\gamma}_0) \equiv \tau_{22} - \tau_{33} \tag{2-37}$$

The normal stress difference growth functions can also be expressed in terms of normal stress growth coefficients.

$$\Psi_1^+(t, \dot{\gamma}_0) \equiv N_1^+/\dot{\gamma}_0^2 \tag{2-38}$$
$$\Psi_2^+(t, \dot{\gamma}_0) \equiv N_2^+/\dot{\gamma}_0^2 \tag{2-39}$$

At steady state, the deformation pattern becomes a motion with constant stretch history, and the values of the viscoelasticity functions become equal to those of the viscometric functions. Of course, the time it takes to achieve a steady state depends on the longest relaxation time and varies from one material to another, so we will use the mathematical limiting case, $t \rightarrow \infty$, to indicate that steady state has been reached.

$$\lim_{t \to \infty} \eta^+(t, \dot{\gamma}_0) = \eta(\dot{\gamma}_0) \tag{2-40}$$

$$\lim_{t \to \infty} N_1^+(t, \dot{\gamma}_0) = N_1(\dot{\gamma}_0) \tag{2-41}$$

$$\lim_{t \to \infty} N_2^+(t, \dot{\gamma}_0) = N_2(\dot{\gamma}_0) \tag{2-42}$$

Or, if the normal stress coefficients are used, Equations 2-43 and 2-44 apply.

$$\lim_{t\to\infty} \Psi_1^+(t, \dot{\gamma}_0) = \Psi_1(\dot{\gamma}_0) \tag{2-43}$$

$$\lim_{t\to\infty} \Psi_2^+(t, \dot{\gamma}_0) = \Psi_2(\dot{\gamma}_0) \tag{2-44}$$

If the shear rate is sufficiently low so that linear viscoelasticity is exhibited, it can be shown that the shear stress growth function becomes independent of shear rate and is directly related to the linear relaxation modulus as follows.

$$\lim_{\dot{\gamma}_0\to 0} \eta^+(t, \dot{\gamma}_0) = \eta^+(t) = \int_0^t G(s)ds \tag{2-45}$$

At larger shear rates, polymeric liquids usually exhibit "stress overshoot"; i.e., both the shear stress and the first normal stress difference go through maxima before approaching steady state values. These maxima are related to the fact that at the same time that the material is responding viscoelastically, the state of the molecules is changing in response to the large shear rate. In dilute solutions, the conformation and orientation distributions are changing, while in melts, the entanglement density is also changing. These changes take time, and they occur over a time scale somewhat longer than that associated with the linear viscoelastic response. Thus, the viscoelastic properties of the material are changing as the experiment proceeds, and the stresses must adapt to this change.

If we suddenly stop the motion after steady state has been achieved in a stress growth experiment, for example, by locking the upper plate in place in a simple shear apparatus, the stresses will decay, eventually becoming isotropic; i.e., all the extra stresses will settle back to zero. This experiment is called "stress relaxation after cessation of steady shear," and the data are usually interpreted in terms of the following stress relaxation functions.

$$\eta^-(t, \dot{\gamma}_0) \equiv \tau/\dot{\gamma}_0 \tag{2-46}$$

$$N_1^-(t, \dot{\gamma}_0) \equiv (\tau_{11} - \tau_{22})/\dot{\gamma}_0 \tag{2-47}$$

$$N_2^-(t, \dot{\gamma}_0) \equiv (\tau_{22} - \tau_{33})/\dot{\gamma}_0 \tag{2-48}$$

The corresponding normal stress coefficient relaxation functions, $\Psi_1^-(t, \dot{\gamma}_0)$ and $\Psi_2^-(t, \dot{\gamma}_0)$ can be used in place of N_1^- and N_2^-.

If the initial shear rate, $\dot{\gamma}_0$, is sufficiently low that linear viscoelasticity is exhibited, then the relaxation functions defined above become independent of the shear rate. In this case, it can be shown that

$$\eta^-(t) = \int_t^\infty G(s)\,ds. \tag{2-49}$$

Stress growth and relaxation curves typical of polymeric liquids are sketched in Figure 2-8. Also shown for comparison is the behavior of a Newtonian fluid. The normal stress differences relax somewhat more slowly than the shear stress. All the stresses relax more quickly as $\dot{\gamma}_0$ is increased.

2.4.3 Shear Creep and Creep Recovery

Another simple transient experiment based on a simple shear flow is the "creep" experiment in which, at the start, the undeformed sample is suddenly subjected to a constant shearing stress, τ_0. The shear strain

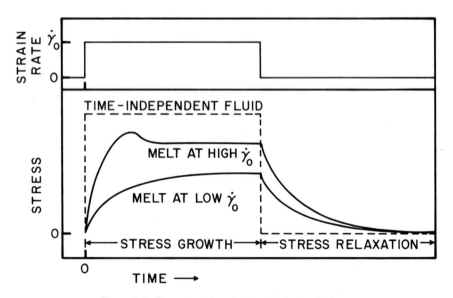

Figure 2-8. Stress growth and relaxation in simple shear.

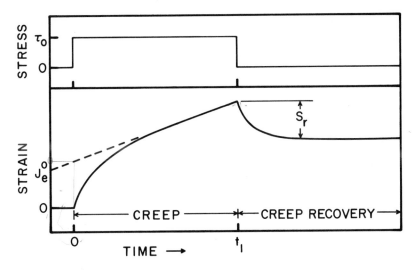

Figure 2-9. Creep and creep recovery curves.

will increase with time, approaching a steady state in which the shear rate is constant, as shown in Figure 2-9. The data are interpreted in terms of the "creep compliance" defined as follows.

$$J(t, \tau_0) \equiv \gamma/\tau_0 \qquad (2\text{-}50)$$

After steady state has been achieved, the strain and, thus, the creep compliance, become linear with time. If this linear portion of the compliance curve is extrapolated back to zero time, the intercept is called the steady state compliance, $J_e^0(\tau_0)$.

In the linear portion of the curve, we have a motion with constant stretch history, for which the shear rate is related to the shear stress through the viscosity.

$$\frac{d\gamma}{dt} \equiv \dot{\gamma} = \tau_0/\eta$$

Thus, the equation for this linear portion of the curve is

$$J(t, \tau_0) = J_e^0(\tau_0) + t/\eta. \qquad (2\text{-}51)$$

If the applied stress is sufficiently small, the creep compliance becomes independent of stress, and the steady state compliance becomes equal to a material constant, J_e^0. In this case, we are measuring the linear viscoelastic material function $J(t)$ and, in the steady state, this becomes

$$J(t) = J_e^0 + t/\eta_0. \tag{2-52}$$

After steady state is achieved in a simple shear experiment, one can observe an explicit manifestation of the liquid's elasticity by suddenly eliminating the shear stress at some time, t_1, so that the upper plate in Figure 1-4 is free to move in the x_1 direction, while keeping constant the spacing between the plates. If the fluid is viscoelastic, it will tend to spring back somewhat, and this phenomenon is called "creep recovery" or "constrained recoil." The material function measured is the strain recovery as a function of time.

$$\gamma_r(t) \equiv \gamma(t_1) - \gamma(t) \tag{2-53}$$

If the initial stress, τ_0, is sufficiently small that we are in the range of linear viscoelasticity, this function is directly related to the creep compliance function.

$$\gamma_r(t - t_1) = \tau_0 \left[J(t - t_1) - \frac{(t - t_1)}{\eta_0} \right]$$

For large values of $(t - t_1)$, Equation 2-52 tells us that

$$J(t - t_1) = J_e^0 + (t - t_1)/\eta_0$$

so that the recoil function approaches a maximum value, S_r, called the "recoverable shear."

$$S_r \equiv \gamma_r(\infty) = \tau_0 J_e^0 \tag{2-54}$$

A "shear modulus," G, is sometimes used, and it is defined as the reciprocal of the steady state compliance, so that Equation 2-54 can also be written as follows.

$$\tau = GS_r \qquad\qquad (2\text{-}55)$$

This equation is called "Hooke's law in shear." Since it is based on Linear viscoelasticity, it cannot be relied upon except at very low shear rates, but it is often observed that it continues to fit data even at shear rates at which the viscosity is dependent on shear rate, implying that the shear modulus is not a strong function of shear rate.

The theory of slow fluid motions, mentioned in the presentation of the viscoelastic functions earlier in this chapter, predicts the following relationship between the steady state compliance, J_e^0, and the limiting low shear rate values of two of the viscometric functions:

$$J_e^0 = \Psi_{1,0}/2\eta_0^2. \qquad\qquad (2\text{-}56)$$

If this is combined with Equation 2-54, the recoverable shear can be related to the limiting values of the viscometric functions.

$$S_r = \tau_0 \Psi_{1,0}/2\eta_0^2 \qquad\qquad (2\text{-}57\text{a})$$

This can also be written as follows.

$$S_r = N_1/2\tau_0 \qquad\qquad (2\text{-}57\text{b})$$

Although this equation is derived from the theory of slow motion,* it has been sometimes claimed that it continues to fit experimental data at shear rates sufficiently large that η and Ψ_1 are dependent on shear rate. For this reason, the quantity on the right-hand side of Equation 2-57b is sometimes called the "recoverable shear." It is also sometimes used as the definition of the "Weissenberg number," which is a ratio of elastic to viscous forces in simple shear flow. Yet a third name for this quantity is the "stress ratio," and if one wishes to minimize the possibility of misunderstanding, this is the name that should be used. Additional relationships between recoverable shear and other rheological properties have been described by Stratton and Butcher [27].

---

*The theory of the so-called second-order fluid.

2.4.4 Small Amplitude Oscillatory Shear

All the test modes discussed so far involve subjecting the material to a step change in shear rate (or shear stress) and measuring the stress (or strain) as a function of time. In the terminology of process dynamics, this would be called a characterization "in the time domain." An alternative procedure which has proved quite useful in the study of polymeric liquids is to subject the material to a periodic deformation. The type of deformation usually used is a sinusoidal simple shear such as might be generated by moving the upper plate in our idealized simple shear experiment back and forth.

The strain as a function of time is given by

$$\gamma = \gamma_0 \sin(\omega t). \tag{2-58}$$

The shear rate is also periodic, and it is given by

$$\dot{\gamma} = \gamma_0 \omega \cos(\omega t) = \dot{\gamma}_0 \cos(\omega t). \tag{2-59}$$

This test is usually used to study the linear viscoelastic properties of polymeric materials. If the strain amplitude, γ_0, is sufficiently small, the shear stress can be written as

$$\tau = \tau_0 \sin(\omega t + \delta) \tag{2-60}$$

where τ_0 is the amplitude of the shear stress and δ is the phase shift relative to the strain, sometimes called the "mechanical loss angle." Furthermore, the dependence of τ/γ_0 on time is independent of γ_0. Thus, if we rewrite Equation 2-60 as Equation 2-61,

$$\frac{\tau}{\gamma_0} = \frac{\tau_0}{\gamma_0} \sin(\omega t + \delta), \tag{2-61}$$

we can conclude that, as long as γ_0 is sufficiently small, the amplitude ratio, (τ_0/γ_0), and the phase shift, δ, are independent of amplitude. This means that the results of a small amplitude oscillatory shear test can be described completely by plots of (τ_0/γ_0) and δ as functions of frequency.

In fact, these particular functions are not the ones commonly used to present the results of a small amplitude oscillatory shear test. It is customary to use these data to calculate the values of material functions defined in such a way that they have a more obvious physical significance. For example, we can rewrite Equation 2-60 as a sum of an in-phase term and a 90° out-of-phase term.

$$\tau = \gamma_0[G'\sin(\omega t) + G''\cos(\omega t)] \qquad (2\text{-}62)$$

The quantities G' and G'' are called the "storage modulus" and "loss modulus," respectively, and they are functions of frequency. Using trigonometric identities, it is easy to find out how to calculate G' and G'' at a given frequency from the amplitude ratio and the phase shift.

$$G' = (\tau_0/\gamma_0)\cos\delta \qquad (2\text{-}63)$$
$$G'' = (\tau_0/\gamma_0)\sin\delta \qquad (2\text{-}64)$$

Alternatively, Equation 2-62 can be written in yet another form:

$$\tau = \dot{\gamma}_0[\eta'\cos(\omega t) + \eta''\sin(\omega t)]. \qquad (2\text{-}65)$$

The two material functions so defined are simply related to the storage and loss moduli.

$$\eta' = G''/\omega = (\tau_0/\dot{\gamma}_0)\sin\delta \qquad (2\text{-}66)$$
$$\eta'' = G'/\omega = (\tau_0/\dot{\gamma}_0)\cos\delta \qquad (2\text{-}67)$$

The function η' is called the dynamic viscosity. The "absolute magnitude of the complex viscosity" is defined as follows.

$$|\eta^*| = \sqrt{(\eta')^2 + (\eta'')^2} \qquad (2\text{-}68)$$

Another material function sometimes used to characterize viscoelastic materials is the tangent of the mechanical loss angle, "tan delta," as a function of frequency. From the above equations, we can see how this is related to the components of the complex modulus and complex viscosity.

$$\tan \delta = \frac{G''}{G'} = \frac{\eta'}{\eta''} \qquad (2\text{-}69)$$

To close this discussion of the use of oscillatory shear to determine linear viscoelasticity properties, let us see what kind of limiting behavior we might expect at very low frequencies. First, we note that when both the amplitude and frequency become very small, a viscoelastic liquid will behave more and more like a Newtonian fluid, because both the extent and the rate of deformation are very small. Thus, the following limiting behavior is expected as the frequency approaches zero.

$$\lim_{\omega \to 0} \eta' = \eta_0 \tag{2-70}$$

$$\lim_{\omega \to 0} \delta = 90° \tag{2-71}$$

$$\lim_{\omega \to 0} G' = 0 \tag{2-72}$$

Furthermore, the theory of very slow motions, mentioned in connection with the definitions of the normal stress coefficients, suggests that the storage modulus and the first normal stress coefficient should be related as follows.

$$\lim_{\omega \to 0} \left(\frac{2G'}{\omega^2} \right) = \lim_{\dot{\gamma} \to 0} \left(\frac{N_1}{\dot{\gamma}^2} \right) \equiv \Psi_{1,0} \tag{2-73}$$

This result implies that a plot of log G' versus log ω should approach a straight line with a slope of two at sufficiently low frequencies. However, such behavior may not be observed at the lowest frequencies at which reliable measurements can be made. This is especially true of polymers having a broad molecular weight distribution.

2.4.5 Superposed Steady and Oscillatory Shear

We have observed that while viscometric flow provides no direct information regarding a fluid's elasticity, small amplitude oscillatory shear provides no information about the effect of large strain on the material properties.* This suggests the possibility of superposing these two types of deformation such that the shear rate is given by

$$\dot{\gamma}(t) = \dot{\gamma}_m + \dot{\gamma}_0 \cos\omega t. \tag{2-74}$$

---

*A partial exception to this statement is the similarity sometimes observed between curves of $|\eta^*|$ as a function of frequency, and viscosity as a function of shear rate. This was first noted by Cox and Merz [28]. The resemblance between the two functions is not observed for all polymers. For example, it is generally not observed in the case of relatively stiff molecules.

Thus, the strain rate oscillates about a mean value, $\dot{\gamma}_m$. If the amplitude of the oscillatory component of the deformation is sufficiently small, the shear stress will also be a harmonic function of time added to a mean value.

$$\tau(t) = \tau_m + \tau_0 \sin(\omega t + \delta) \qquad (2\text{-}75)$$

The mean stress is given by

$$\tau_m = \eta(\dot{\gamma}_m)\dot{\gamma}_m \qquad (2\text{-}76)$$

and the time-dependent component can be written as follows.

$$\tau - \tau_m = \dot{\gamma}_0[\eta'\cos(\omega t) + \eta''\sin(\omega t)] \qquad (2\text{-}77)$$

In this case, however, the two material functions, η' and η'', depend on both the frequency and the mean shear rate. The usual procedure is to plot these functions versus frequency with $\dot{\gamma}_m$ as a parameter.

Another possibility is to carry out the oscillating shear in a plane perpendicular to that of the steady shear to obtain yet another rheological characteristic of the fluid.

Bird et al. ([1], p. 167) have reviewed the use of these superposed flows in the study of polymeric liquids.

2.4.6 Large amplitude oscillatory shear

Another way of studying the elastic properties of a material outside the realm of linear viscoelasticity is to measure the stress resulting from an oscillatory shear like that given by Equation 2-58, but with a strain amplitude sufficiently large that the material structure is altered.

Unfortunately, the data from such an experiment cannot be reduced to two material functions such as $\eta'(\omega)$ and $\eta''(\omega)$, as the function (τ/γ_0) is neither harmonic nor independent of amplitude. Tee and Dealy [29] have discussed the problems that arise in the interpretation of data from this type of experiment. They suggest that, for purposes of comparison, a plot of stress versus strain rate be prepared. Such a plot appears as a closed loop, and in the limiting case of small amplitude, the loop is elliptical in shape, as shown in Figure 2-10a. The deviation in shape from an ellipse at larger amplitudes is characteristic of a non-

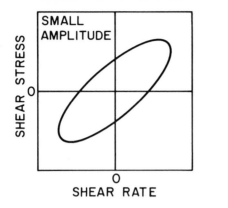 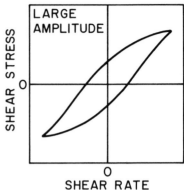

Figure 2-10. Stress versus rate-of-strain loops for oscillatory shear: (a) small amplitude, (b) large amplitude (after Tee and Dealy [29]).

linear response. An example of the type of loop observed by Tee and Dealy [29] for a branched polyethylene at large amplitude is shown in Figure 2-10b.

2.4.7 Transient Extensional Flows

Each of the transient simple shear flows described in the preceding sections has an analogous simple extensional flow. For example, one can define a relaxation modulus in extension as follows.

$$E(t, \epsilon_0) = \frac{\tau_{11} - \tau_{22}}{\epsilon_0} \tag{2-78}$$

If the strain is sufficiently small, the modulus is independent of strain and is closely related to the shear modulus as follows.

$$E(s) = 3G(s) \tag{2-79}$$

Thus, within the realm of linear viscoelasticity, we learn nothing new by performing stress relaxation in simple extension if we already know $G(s)$.

The creep compliance in simple extension is

$$D(t, \tau_{11} - \tau_{22}) \equiv \epsilon(t)/(\tau_{11} - \tau_{22}). \tag{2-80}$$

If the stretching stress is sufficiently small, the compliance is independent of stress, and is equal to one-third of the shear compliance. The steady state extensional compliance is thus related to the steady state shear compliance as follows.

$$D_e^0 = J_e^0/3 \qquad (2\text{-}81)$$

The stress growth and relaxation functions for steady simple extension are defined as follows.

$$\eta_T^+(t, \dot{\epsilon}_0) = (\tau_{11} - \tau_{22})/\dot{\epsilon}_0 \qquad (2\text{-}82)$$
$$\eta_T^-(t, \dot{\epsilon}_0) = (\tau_{11} - \tau_{22})/\dot{\epsilon}_0 \qquad (2\text{-}83)$$

If $\dot{\epsilon}_0$ is sufficiently small, it can be shown that these functions are closely related to the corresponding shear functions.

$$\eta_T^+(t) = 3\eta^+(t) \qquad (2\text{-}84)$$
$$\eta_T^-(t) = 3\eta^-(t) \qquad (2\text{-}85)$$

Of course, the corresponding time-dependent material functions can be defined for biaxial and planar extensions.

Chapter 3
Basic Problems in the Design and Use of Melt Rheometers

Before proceeding to a detailed description of rheometrical flows and geometries, it will be useful to enumerate the basic experimental problems that arise in the use of all of them. Every rheometer has its limitations, and it is essential to understand these if reliable results are to be obtained.

3.1 DEVIATIONS FROM THE PRESCRIBED DEFORMATION

As was pointed out at the beginning of Chapter 2, if we wish to measure a well-defined material property, we must use an apparatus that always generates the same flow regardless of the fluid studied. A flow in which the velocity field is governed only by the boundary conditions regardless of the fluid's rheological properties, is called a "controllable flow" [30]. Simple shear between parallel plates is a controllable flow, but it is not a practical basis for rheometer design. Therefore, rheometers that measure well-defined material functions are designed around flows that are either "approximately" or "partially" controllable [30].

An approximately controllable flow is one in which controllability is achieved only to the extent that certain terms, usually acceleration terms, in the equations of motion can be neglected. An example is cone-plate flow. However, the neglect of these terms is only valid at low deformation rates.

A partially controllable flow is one in which the streamlines depend only on the boundary geometry, but in which the velocity distribution depends on the fluid properties. An example is flow in a circular tube. The use of partially controllable flows to determine rheological properties requires the use of special techniques for analyzing the data so that the deformation rate can be inferred.

3.1.1 Flow Instability

A particular flow geometry, such as concentric rotating cylinders or a capillary tube, which is useful for generating a simple controllable flow over certain ranges of fluid properties, apparatus dimensions and speed, may generate a more complex, uncontrollable flow under conditions outside these ranges. The problem of determining which of several possible flows is the one most likely to be observed is called a "hydrodynamic stability" problem.

The controllable flow assumed in the derivation of rheometrical equations is always the simplest possible flow for a given geometry, and if this flow becomes unstable, the result may be another, more complex flow involving secondary motions superposed on the simplest or primary flow. Or it may be a turbulent flow in which there are random variations in velocity.

In the case of Newtonian fluids, such instabilities have been studied extensively in the laboratory and can often be predicted theoretically. For example, in the flow of a given fluid between rotating concentric cylinders, if the inner cylinder is rotated above a certain speed, regularly spaced toroidal vortices appear. If the speed is increased further, a point is reached where the motion becomes turbulent. For a Newtonian fluid flowing in a tube, it is well known that if the Reynolds number* exceeds 2,100, a slight disturbance can trigger a transition to turbulence.

Whenever a secondary flow or turbulence occurs, rheometrical equations based on the simple controllable flow are no longer valid, and their use will lead to the calculation of incorrect values for rheological properties. Thus, it would be advantageous to be able to predict the occurrence of hydrodynamic instability for the flows in melt rheometers.

While it is often possible to predict the onset of secondary flows and turbulence for Newtonian fluids, in the case of viscoelastic fluids, it has not been possible to establish general criteria for the stability of various rheometrical flows. The high viscosity of molten polymers might be expected to enhance the stability of the simple, low-speed flow pattern, but the effect of fluid elasticity on the stability is not well understood, so that the melt rheologist must be always alert to the possible occurrence of flow instabilities.

---

*The Reynolds number is the product of the tube diameter and the average flow velocity, divided by the kinematic viscosity.

3.1.2 Edge and End Effects

Rheometrical equations that are used to calculate rheological properties from raw experimental data are derived on the basis of a certain assumed flow pattern, but there is always a region of the actual rheometer in which the flow deviates from this idealized pattern. Examples are the flow near the entrance of a tube, the flow near the ends of a concentric cylinder rheometer and the flow at the edge of a cone-plate rheometer. Errors resulting from these end and edge effects can be minimized by proper design of the rheometer. In some cases, as we shall see, their effect on the measurement of properties can be eliminated by treating the data in a special way.

3.1.3 Melt Fracture

When an entangled polymer melt is subjected to large, rapid increases in deformation rate, the adaptation of the entanglement density to the new situation may occur sufficiently slowly that the material response is basically that of a rubber and the melt can undergo a fracture process. Such a situation can arise at the entrance to a tube or in large amplitude oscillatory shear, and Vinogradov [31] has described in some detail the fracture of melts in these situations. Obviously, when fracture occurs, meaningful rheological data cannot be obtained.

3.1.4 Variation of Rheometer Geometry

In order to minimize the effects of fluid inertia and promote temperature uniformity, rheometer geometries are normally designed to have rather small clearances between the solid bounding surfaces. This makes it quite important that these bounding surfaces be manufactured to close tolerances and that they maintain their positions and orientations during operation of the rheometer. Rheometer manufacturers usually provide detailed specifications with the fixtures they sell, but if highly reliable results are required, the fixture geometry should be verified by the user.

Once we have installed the fixtures that provide the bounding surfaces for the flow in the rheometer and verified their orientation and spacing, it is possible for the flow geometry to change during the use of the rheometer. For example, the elevated temperatures normally used in melt studies will cause thermal expansion of various rheometer components, and this will result in dimensional changes. Furthermore, the rheometer structure will deform in response to the forces applied to the melt, and this structure must be designed to be sufficiently stiff (i.e., to have a sufficiently low compliance) so that its elastic deformation does not significantly alter the flow geometry.

3.1.5 Speed and Position Error

Unless the deforming force is supplied by a suspended weight or gas pressure, some kind of motor will be required to generate the desired motion, and it is essential that the speed of the motor be known precisely in order to calculate the strain rate. One way of generating motion at a precisely known speed is by use of a synchronous motor. However, the speed cannot be varied continuously, and the more common procedure is to employ a feedback control loop to control the motor speed and maintain it at a preset level. Analog control circuits are subject to drift, and repeated checks of the speed constancy and calibration may be necessary. This problem is especially severe when operation at very low speeds is desired, for example, to determine η_0. Digital controllers provide more precise control of position or speed and are especially advantageous for low-speed operation or when on-line data acquisition and analysis is to be employed.

Of course, a motor and the mechanical assembly to which it is directly coupled cannot change speed instantaneously because of inertia and frequency response limitations. This means that it is not possible to generate with exactitude the strain histories that serve as bases for some of the material functions defined in Chapter 2; for example, the stress growth function. However, if instrument inertia is minimized and a good quality control system is employed, reliable results can be obtained.

3.2 PRESSURE HOLE ERROR

Some rheometrical methods require the measurement of the component of stress normal to a solid wall past which the liquid is flowing. In order to minimize the disturbance to the flow, it was for some years thought that this could best be accomplished by mounting a pressure transducer in a chamber communicating with the flowing liquid by means of a small hole, as shown in Figure 3-1. It was found that if the hole radius was less than a certain value, the measured pressure was independent of hole size, and it was inferred from this that the hole had no effect on the flow and that the pressure measured in the stagnant liquid in the chamber was equal to the normal stress at the wall.

However, comparative tests using more than one measurement technique [32] revealed significant disagreement leading to the suggestion [33] that there was an intrinsic error involved in the use of "pressure holes." Tanner and Pipkin [34] observed that, even with a small hole, the streamlines tend to dip into the hole, and that, in the case of a viscoelastic fluid with a positive first normal stress difference, this might be expected to result in the measured pressure, P_m, being less than the true wall pressure, P_w, in the undisturbed flow.

Han [35] has suggested that the pressure hole error for melts is quite small, but data currently available are insufficient to permit one to state this as a general rule. Of course, if one uses two pressure holes to measure ΔP_m in a fully developed rectilinear flow, then the flow is the same over both holes, and

$$\Delta P_w = \Delta P_m \qquad (3\text{-}1)$$

In Section 4.5.3 is described a technique which has been proposed to use the pressure hole error as a basis for estimating the first normal stress difference.

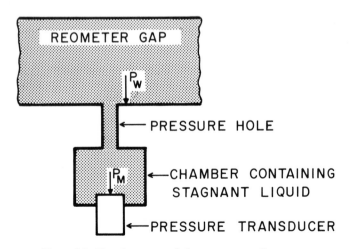

Figure 3-1. Use of a pressure hole to measure wall pressure.

3.3 TEMPERATURE NONUNIFORMITY

Rheological properties are strong functions of temperature, so that the control of this variable is of central importance. The control of temperature in a melt rheometer is complicated by the need to maintain the sample at temperatures from $100°C$ to $300°C$ above room temperature.

The maintenance of a uniform temperature poses special problems in the case of polymer melts, because the high viscosity leads to a high rate of viscous heating. Whenever a viscous liquid is deformed, some of the work of deformation is converted to internal energy through the mechanism of viscous dissipation, and this results in a temperature increase. In order to maintain the temperature at a prescribed value, this energy must be removed as heat. However, heat only flows when there is a temperature gradient. As a result, it is impossible to maintain a perfectly uniform temperature in a viscous material while it is being deformed. While we can design our experiment in such a way as to keep the maximum temperature variation within acceptable bounds, we can never eliminate it entirely.

As an aid to the understanding of viscous heating, it is useful to consider the very simple case of uniform shearing of a Newtonian fluid whose viscosity is independent of temperature. If both bounding planes are maintained at a fixed temperature, T_w, the steady state temperature distribution is as follows.*

$$T(y) = T_w + \frac{\eta \dot{\gamma}^2 h^2}{2k} \left[\frac{y}{h} - \left(\frac{y}{h} \right)^2 \right] \qquad (3\text{-}2)$$

This distribution is shown in the lower curve of Figure 3-2. The maximum temperature occurs at the midplane where $y = h/2$ and is given by Equation 3-3.

$$T_{max} = T_w + \frac{\eta \dot{\gamma}^2 h^2}{8k} \qquad (3\text{-}3)$$

Thus, the temperature variation within the gap between the planes is

$$T_{max} - T_w = \frac{\eta \dot{\gamma}^2 h^2}{8k}. \qquad (3\text{-}4)$$

---

*The details of this analysis can be found on p. 10 of [1].

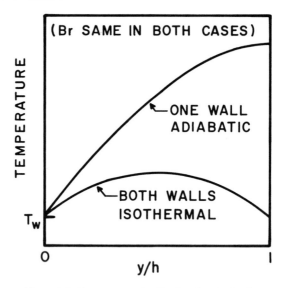

Figure 3-2. Temperature distributions for simple shear.

If only one of the two bounding planes is maintained at the control temperature, T_w, while the other surface is insulated so as to eliminate any heat flow through it, the temperature distribution is altered as shown in the upper curve in Figure 3-2. In this case, the maximum temperature occurs at the insulated surface, and the temperature variation is as follows.

$$T_{max} - T_w = \frac{\eta \dot{\gamma}^2 h^2}{2k} \tag{3-5}$$

Thus, the temperature variation within the fluid is four times as great as in the case when both walls are maintained at the control temperature.

If we take into account the variation of viscosity with temperature, the analysis of viscous heating becomes more complex, but certain general observations can be made that will still be valid in this more realistic case. First, the extent of the temperature nonuniformity increases markedly as the shear rate or gap spacing is increased. It also increases with viscosity, but less sharply, and it decreases as the thermal conductivity is increased. Finally, the nonuniformity is much less when both plates are maintained at the control temperature than when one plate is insulated.

The extent to which the temperature variation affects a property measurement depends on the sensitivity of that property to temperature. For example, let us say that we wish to measure viscosity, and that the viscosity varies with temperature as follows.

$$\eta(T) = \eta(T_w)e^{-b(T-T_w/T_w)} \tag{3-6}$$

Another way of expressing this is

$$\ln\left[\frac{\eta(T_w)}{\eta(T)}\right] = b\left(\frac{T-T_w}{T_w}\right).$$

If the viscosity is a monotonic function of temperature, the ratio of the maximum to minimum values of the viscosity in the gap is given by

$$\ln\left[\frac{\eta(T_w)}{\eta(T_{max})}\right] = \ln\left[\frac{\eta_{max}}{\eta_{min}}\right] = b\left(\frac{T_{max}-T_w}{T_w}\right)$$

and, for the case in which both walls are at T_w, this can be combined with Equation 3-4 to give

$$\ln\left[\frac{\eta_{max}}{\eta_{min}}\right] = \frac{b\eta\dot{\gamma}^2 h^2}{8kT_w}. \tag{3-7}$$

For the case in which one wall is insulated, we have:

$$\ln\left[\frac{\eta_{max}}{\eta_{min}}\right] = \frac{b\eta\dot{\gamma}^2 h^2}{2kT_w}. \tag{3-8}$$

If we wish to limit the variation in viscosity to 5% ($\eta_{max}/\eta_{min} = 1.05$), the product $\dfrac{b\eta\dot{\gamma}^2 h^2}{kT_w}$ must be less than 0.4 for the case of two isothermal boundaries and less than 0.1 when one boundary is adiabatic. If the fluid is non-Newtonian with temperature-dependent material constants, and if the geometry is more complex, these guidelines are not quantitatively valid, but they nonetheless expose the role played by the various experimental variables and fluid properties.

One approach to the problem of viscous heating is to make measurements before there has been sufficient time for the significant accumulation of viscous heat. For this technique to be feasible, the buildup of stress after the start of the deformations must occur much more rapidly than the development of the temperature distribution. In the case of a Newtonian fluid, this will occur when the Prandtl number is high.* Figure 3-3 shows the influence of viscous heating on the buildup of the stress at the start of a steady strain-rate experiment.

The upper curve in Figure 3-3 is for the case of negligible viscous heating, and the lower curve corresponds to substantial viscous heating with a fairly high Prandtl number so that the maximum in the stress curve is very close to the steady state value that would be observed in the absence of viscous heating.

Of course, it is essential that the temperature probe used to measure temperature and to provide that the feedback signal for a temperature controller be located so that its temperature is as close as possible to that of the melt.

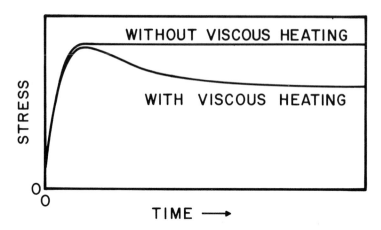

Figure 3-3. Variation of stress with time at startup of simple shear with and without viscous heating.

---

*The Prandtl number is the product of viscosity and heat capacity divided by the product of the density and the thermal conductivity.

3.4   EFFECTS OF PRESSURE AND ENVIRONMENT

As has been explained in Chapter 1, the term "pressure" has no single, unambiguous operational meaning in the case of melts undergoing deformation, and this term is used here to refer to the general level of the normal stresses, σ_{ii}. If we alter the level of all normal forces while keeping the strain rate and geometry constant, we find that it has an effect on the rheological properties, although the effect is not as strong as that of temperature.

The effect of "pressure" or normal stress level on rheological properties poses two types of experimental problems. First, there is the problem of interpreting the results of experiments in which a large "pressure" gradient cannot be avoided. This situation arises when slit or tube rheometers are used to study high viscosity materials at high shear rates, and methods of dealing with this will be described in Chapter 4. The second problem arises when we wish to study the effect of pressure on rheological properties. If the flow of interest is steady shear at high strain rate, then a pressure driven flow can be used, but the interpretation of the data is complex and does not lead to completely unambiguous results.

It is more attractive to be able to carry out the flow at various ambient pressures without introducing a normal stress gradient in the melt.* This can be done by placing the rheometer fixtures in a sealed chamber filled with gas or oil at a controlled temperature and pressure, although there is now the problem of introducing the deforming force or torque.

Now we turn to the question of the chemical environment. At the high temperatures normally used for melt studies, thermal degradation of the polymer is inevitable, and this limits the useful life of any one sample. Frequent reproduction of experiments is the only reliable technique for ensuring that a sample has not changed significantly, and it is also useful in the detection of other sources of error such as temperature changes and the drift of electronic signals.

Commercial resins nearly always contain stabilizers that retard degradation. Melting under a vacuum and operation in a nitrogen atmosphere will provide further retardation. This is especially important in

---

*Of course, there is always a "hydrostatic" or gravity induced gradient of normal force, but this is negligible in the case of polymer melts.

the case of polymers such as nylons, polyesters and polycarbonates, which are highly sensitive to moisture at melt temperatures.

3.5 LOADING AND CLEANING

Unless the polymer under study is molten at room temperature, one cannot load a rheometer by pouring liquid into place. Except in pressure-driven tube and slit rheometers, loading with resin pellets is likewise impractical because it is nearly impossible to control the sample volume and eliminate bubbles and voids. For these reasons, one usually uses a premolded sample that has, at least approximately, the shape of the test geometry to be used. Transfer, compression or injection molding, and even extrusion can be used to prepare samples. In the case of shear flows, the rheometer fixtures will maintain the shape of the sample during operation, but in extensional flows, most of the sample surface is not in contact with solid boundaries, so special care must be taken to prepare premolded samples which are strain-free and have exactly the shape and size required.

Cleaning of a melt rheometer is always a nuisance, and care must be taken to avoid damage to the rheometer. Melt rheometers must be designed in such a way as to permit good access to all surfaces in contact with the melt. Polymer that is not removed will degrade, and this will reduce the reliability of results and make the ultimate removal of the polymer all the more difficult.

3.6 VERIFICATION OF RESULTS

An essential and often neglected aspect of property measurement is calibration and continuous verification of results. In the case of melt rheometers, this is not a simple matter. First, by measuring fixture dimensions and calibrating force transducers, speed or displacement sensors and temperature probes, one can develop confidence in each component of the rheometer. However, it is always wise to perform a periodic calibration involving the actual operation of the rheometer with a molten polymer. Such a calibration requires the availability of either a well-characterized standard reference material or the facilities to compare data from more than one rheometer.

Standard Newtonian calibration standards are available commercially,* but the most viscous standard fluids are much less viscous than typical thermoplastic melts. Furthermore, they provide a calibration only for viscosity and not for other viscoelastic properties.

A useful procedure is to compare the data on a single resin obtained from two or more rheometers. Since few laboratories are so well-equipped as to have more than one instrument for measurement of a given property, this may require cooperation between two or more laboratories.

Once a rheometer has been calibrated, it is still wise to check the reproducibility of data and perform other periodic verification experiments. It is the obligation of the supervisor of property testing procedures to ensure the precision of the test results. Manufacturers' specifications and calibrating data must not be looked upon as absolute truth, the correctness of which need not be questioned.

*Cannon Instrument Co., State College, Pa.

Chapter 4
Rheometrical Flows Involving Rectilinear Shear

The simplest approach to the use of shearing flows in rheometry is to employ a flow geometry that generates rectilinear flow with a velocity gradient. In such a flow, all the streamlines are straight and parallel. Rectilinear flow geometries which have been used to study melts include sliding plates and sliding cylinders, in which the shear is generated by moving one wall of the rheometer, as well as capillaries and slits, in which the shear results from a pressure-driven flow through a conduit. Sliding plate flow is a controllable flow, except for edge and end effects, while the remaining geometries mentioned are only partially controllable.

4.1 SLIDING PLATE RHEOMETERS

The laboratory procedure that most closely approximates simple shear is to place a thin layer of liquid between two flat plates, clamp one of the plates in place, and translate the second plate at a constant velocity, as shown in Figure 4-1. The shear rate is given by Equation 1-6 and the viscosity by Equation 1-9. An advantage of the sliding plate geometry over other geometries used to carry out shearing tests is that it can be used to study orientation effects (for example, in glass fiber reinforced plastics).

This technique has been applied to the study of viscous, Newtonian liquids, and in this case the fluid itself serves to maintain the plate spacing. However, if the first normal stress difference is positive, which seems to be the case for molten polymers, then the shearing deformation will result in a force tending to separate the plates, and some way must be found to maintain the gap without introducing a significant frictional force that would interfere with the determination of the viscosity.

However, unless the second normal stress difference is exactly the negative of the first normal stress difference, there will be a secondary flow with components in the x_1 and x_3 directions. If N_2 is negative, as seems to be the case for at least some melts, the effect of this secondary flow will be to cause fluid to flow out of the gap along the side edges as shown in Figure 4-2.

Finally, we note that as the motion proceeds, the effective shearing area decreases, because the plates are moving away from each other in the direction of flow as shown in Figure 4-3. It is difficult to account for this quantitatively because of the changing orientation of the free surface, and this means that the use of sliding plate rheometers is limited to rather small total strains, perhaps 200% to 300%.

Fruh and Rodriguez [36] used a parallel plate rheometer to study the creep of molten polymers. The driving force was applied by means of a weight suspended from a pulley mounted on an air bearing. Two stationary plates and one moving plate were employed with a layer of sample in each of the two gaps. This arrangement eliminates the tendency of the fluid to displace the moving plate in a direction normal to the flow. The moving plate was supported and linked to the pulley through a linear air bearing, and the motion was detected by means of

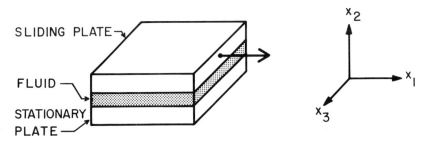

Figure 4-1. Sliding plate geometry.

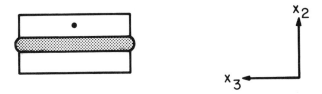

Figure 4-2. End view of sliding plate geometry. (The upper plate is moving toward the reader.)

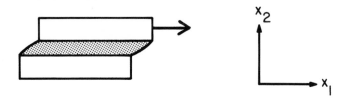

Figure 4-3. Side view of sliding plate geometry showing variation of shearing area with time.

a linear, variable differential transformer. Strain rate and total strain were limited to small values.

Fujiyama and Takayanagi [37] used a plate creepmeter in which the moving plate was mounted vertically and inserted into a rectangular box-like test cell filled with molten polymer. An upward force was applied by means of a weight on the opposite end of a lever.

Laun and Meissner [38] have described a weight-loaded, parallel plate creepmeter making use of a "sandwich" type of fixture. Two square sheets of the material to be tested are placed on either side of a sheet of metal. Two additional metal sheets are placed on the outer sides of the polymer layers. The deforming stress is provided by applying a force to the central plate parallel to its plane and a force in the opposite direction to the two outer plates. The two outer plates are mounted in a fixture that maintains the spacing between them. Laun and Meissner made an estimate of the contribution of surface tension to the total force exerted by the melt on the metal sheets, and found that in the case of their study, this contribution was about 2% of the total. They also note that in creep recovery, surface tension continues to provide a driving force for deformation after elastic recoil is complete.

The sandwich-type mounting had been used previously by Middleman [39], La Mantia et al. [40] and others, particularly for the study of elastomers, as the basis of constant strain rate rheometers. In these devices, the central plate and the outer plates were coupled to the crossheads of a standard tensile testing machine. If the crossheads are driven apart at a constant speed, a stress growth experiment can be carried out.

4.2 SLIDING CYLINDER FLOW

In this flow, sometimes called "telescopic flow," the fluid is contained in the annular gap between a solid cylinder, with radius r_i and length

L, and a hollow cylinder with inside radius r_0, as shown in Figure 4-4. The inner cylinder moves axially, and the force applied to the end of this cylinder and its corresponding velocity of displacement, V, are measured.

The shear stress at the wall of the sliding cylinder is related to F (a positive quantity) as follows.

$$\tau_w \equiv - \tau_{rz}(r_i) = \frac{F}{2\pi r_i L} \qquad (4\text{-}1)$$

The velocity profile for a Newtonian fluid is given by Equation 4-2.

$$v(r) = V\left(\frac{\ln r/r_0}{\ln r_i/r_0}\right) \qquad (4\text{-}2)$$

The shear rate at the wall of the sliding cylinder is

$$\dot{\gamma}_w \equiv -\left(\frac{\partial v}{\partial r}\right)_{r_i} = \frac{V}{r_i \ln(r_0/r_i)}. \qquad (4\text{-}3)$$

The velocity has been taken as positive in the direction of motion so that the shear rate and the shear stress have negative values, and the quantities $\dot{\gamma}_w$ and τ_w are defined, for convenience, so that they have positive values. The viscosity of a Newtonian fluid can then be computed as follows.

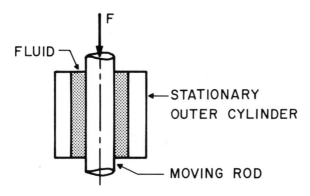

Figure 4-4. Sliding cylinder geometry.

$$\eta = \tau_w / \dot{\gamma}_w = \frac{F \ln(r_0/r_i)}{2\pi LV} \qquad (4\text{-}4)$$

For the case of a power law fluid, as defined by Equation 2-10, the velocity profile is given by

$$v(r) = V \left[\frac{(r/r_0)^{1-1/n} - 1}{(r_i/r_0)^{1-1/n} - 1} \right]$$

from which a procedure for determining the power law parameters, k and n, from experimental data, can be developed.

If a sliding cylinder rheometer is designed so that the relative gap width, E, defined as $(r_0 - r_i)/r_0$, is very small, then the shear rate becomes nearly uniform, and is given approximately by the following simple expression.

$$\dot{\gamma}_w = \frac{V}{(r_0 - r_i)} \qquad (4\text{-}5)$$

This can be recognized as the equation for the shear rate in simple shear, as given by Equation 1-6. Under these circumstances, the variation of viscosity with shear rate will not affect the velocity distribution, and the equation will apply equally to the flow of a non-Newtonian fluid. Thus, for the small gap case, the viscosity of any viscous fluid can be calculated as follows.

$$\eta = \frac{F(r_0 - r_i)}{2\pi r_i LV} \qquad (4\text{-}6)$$

Except in the special case of a small gap, there is no general method for interpreting the data from a sliding cylinder rheometer in terms of viscosity when the form of the viscosity function is unknown. Therefore, this flow has not proven useful when the viscosity is a strong function of shear rate. Also note that the liquid must be sufficiently viscous that gravity will result in no significant downward flow out of the gap during the experiment.

An early report of the use of a sliding cylinder rheometer was that of Pochettino [41], and such devices are sometimes called "Pochettino viscometers." Myers and Faucher [42] used this geometry to measure the viscosity of molten polymers at low shear rates. An Instron testing machine* was used to drive the piston at various speeds, and F was measured by means of a compressive load cell.

McCarthy [43] has studied the dynamic mechanical properties of molten polymers by forcing the inner cylinder to oscillate in the axial direction. This rheometer incorporates a unique sample loading technique that avoids the preparation of a molded preform. A ring-shaped melting chamber surrounding the outer cylinder is charged with resin, and after being melted, it is extruded into the rheometer gap by means of a second ring which is forced down by the manual application of torque. The oscillation of the inner cylinder is driven by an MTS servo-hydraulic ram. The oscillating cylinder is hollow and made from magnesium to minimize its mass. Its temperature is controlled by means of a ribbon heater next to its inner wall.

A modification of the sliding cylinder system, which simplifies sample loading and eliminates the flow of sample out of the gap, is the "falling cylinder rheometer" in which the inner cylinder falls freely as a result of its own weight into an outer cylinder with a closed bottom, as shown in Figure 4-5. In this case, there is an upward flow of liquid in the gap, resulting from the displacement of fluid from the space under the falling cylinder, and the velocity distribution has the form sketched in Figure 4-6. The driving force is

$$F = (\pi r_i^2 L)(\rho_c - \rho_f) \tag{4-7}$$

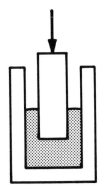

Figure 4-5. Falling cylinder geometry. (The falling cylinder is normally completely submerged)

*See Appendix B for addresses of all manufacturers mentioned in this book.

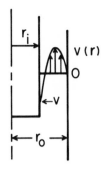

Figure 4-6. Velocity distribution for falling cylinder flow.

where ρ_c and ρ_f are the densities of the cylinder and fluid. The equation for calculating the viscosity of a Newtonian fluid, as given by Smith *et al.* [44], is

$$\eta = \frac{F[(r_0^2 - r_i^2)\ln(r_0/r_i) - (r_0^2 - r_i^2)]}{2\pi LV(r_0^2 + r_i^2)}. \tag{4-8}$$

Ashare *et al.* [45] carried out a theoretical analysis of falling cylinder flow and presented a series expansion for the velocity with the relative gap width, $[(r_0 - r_i)/r_0]$, as the perturbation parameter. They presented results for a power law fluid and one other empirical expression for $\eta(\dot{\gamma})$ and also a general differentiation procedure for use with any non-Newtonian fluid.

Since the driving force is transmitted into the rheometer without the use of any mechanical linkage, falling cylinder rheometers are well-suited to the measurement of viscosity at high pressure. MacLachlen [46] and Ramsteiner [47] designed high pressure, falling cylinder rheometers, and the latter author made use of an end correction to allow for the fact that the liquid does not reach its fully developed velocity profile immediately upon flowing into the annular space from the bottom of the outer cylinder.

Chee and Rudin [48] used a falling cylinder rheometer in which the inner cylinder is not submerged in the liquid. The immersed length, L_e, of the falling cylinder thus varies with time, and this requires a modification of the rheometer equations. For example, the immersed length, L_e, is related to the displacement, H, of the cylinder as follows.

$$L_e = \left(\frac{r_0^2}{r_0^2 - r_i^2} \right) H \qquad (4\text{-}9)$$

Rudin *et al.* [49, 50] have suggested that this type of arrangement could be used to measure viscosity at high shear rates by using a mechanical testing machine to drive the motion of the inner cylinder at fixed speeds.

4.3 CAPILLARY FLOW

The measurement of the flow rate in a tube as a function of pressure drop is the technique that has been most popular for the study of the rheological properties of liquids, because it is the viscometric flow most easily generated in the laboratory. Among rheologists, tube flow is commonly referred to as "capillary flow," because it is desirable to use a small tube radius to minimize the effect of viscous heating and, in the case of low viscosity liquids, to provide a pressure drop of significant magnitude in a tube of convenient length. In the study of molten polymers, capillary rheometers are useful primarily for the measurement of viscosity at high shear rate and, at the present time, only capillary and slit rheometers are suitable for making such measurements. Methods have also been proposed for the use of capillary rheometers to determine recoverable shear and normal stress differences, but these involve the use of highly questionable assumptions in the interpretation of data.

Capillary flow is an example of a partially controllable flow. Far from the entrance where the flow is fully developed, the streamlines are always parallel to the axis of the tube, but the velocity profile depends on the rheological nature of the fluid. Unless a specific constitutive equation is known to be valid for the fluid, as in the case of a Newtonian fluid or a power law fluid, special computational techniques are required to calculate the viscosity. Other limits to the usefulness of capillary rheometers result from viscous heating, dependence of viscosity on pressure, flow instabilities and end effects. Each of these will be discussed in this chapter.

4.3.1 Basic Equations for Capillary Rheometers

For the steady flow of an incompressible fluid in a tube of radius R, the absolute value of the shear stress at the wall, for which we will use the symbol τ_w, is related to the pressure drop, ΔP, over a length of tube, L, as follows.

$$\tau_w \equiv - \tau_{rz}(r = R) = \frac{-\Delta P \cdot R}{2L} \tag{4-10}$$

The pressure drop, ΔP, is always a negative quantity, because the flow is in the direction of the axial coordinate, z.

$$\Delta P = P(z = z_0 + L) - P(z = z_0)$$

Thus, the shear stress at the wall, $\tau_{rz}(R)$, is also negative.

As this is a partially controllable flow, the velocity profile depends on the viscosity function, and we cannot derive a general expression relating the volumetric flow rate, Q, to the wall shear rate. Before discussing a way around this problem, it will be useful to present some equations valid for the special case of a Newtonian fluid.

For a Newtonian fluid, the velocity distribution is given by the familiar parabolic law shown below.

$$v = 2\bar{v}[1 - (r/R)^2] \tag{4-11}$$

This is the velocity profile for "fully developed flow" in which the effects of the entrance and exit are assumed negligible and there is thus no velocity component in the radial direction. The average velocity, $\bar{v}$, is defined in terms of the volumetric flow rate, Q, as follows.

$$\bar{v} = Q/(\pi R^2) \tag{4-12}$$

The absolute magnitude of the shear rate at the wall, $\dot{\gamma}_w$, can be determined from Equation 4-11.

$$\dot{\gamma}_w \equiv - \left(\frac{dv}{dr}\right)_{r=R} = \frac{4\bar{v}}{R} = \frac{4Q}{\pi R^3} \tag{4-13}$$

The definition of the viscosity tells us that

$$\tau_w = \eta \dot{\gamma}_w \tag{4-14}$$

Combining this with Equations 4-10 and 4-13, and solving for the viscosity, we obtain

$$\eta = \frac{(-\Delta P)\pi R^4}{8LQ}. \qquad (4\text{-}15)$$

Turning now to non-Newtonian fluids, if a specific mathematical form for the viscosity function is assumed, one can derive equations analogous to those valid for Newtonian fluids. For example, if the power law given by Equation 2-10 is assumed, it can be shown that the wall shear rate is given by

$$\dot{\gamma}_w = \left(\frac{3n + 1}{4n}\right)\left(\frac{4Q}{\pi R^3}\right). \qquad (4\text{-}16)$$

It is clear from this result that the quantity $(4Q/\pi R^3)$, which is equal to the wall shear rate in the case of a Newtonian fluid, no longer has this significance when the fluid is non-Newtonian. It is, however, often referred to as the "apparent wall shear rate." We will use the symbol $\dot{\gamma}_A$ for this quantity.

$$\dot{\gamma}_A \equiv \frac{4Q}{\pi R^3} \qquad (4\text{-}17)$$

Equation 4-16 can be used to calculate the error involved in using the apparent wall shear rate as an estimate of the true value. For example, when the power law index, n, is 0.5, the actual wall shear rate is 25% greater than the "apparent" value, and when $n = 0.25$, the deviation is 75%.

Combining Equations 4-10, 4-14 and 4-16, it can be shown that

$$\frac{(-\Delta P)R}{2L} = k\left(\frac{3n + 1}{4n}\right)^n \left(\frac{4Q}{\pi R^3}\right)^n. \qquad (4\text{-}18)$$

Thus, if we make a plot of

$$\log\left[\frac{(-\Delta P)R}{2L}\right] \quad \text{versus} \quad \log\left(\frac{4Q}{\pi R^3}\right)$$

we will obtain a straight line, the slope and intercept of which can be used to determine values for k and n.

If an equation relating the viscosity to the shear rate is not available, then we cannot derive expressions for the velocity profile and the shear rate at the wall in terms of material constants. However, there is a way to calculate values for the viscosity, provided that a large number of data are available over a range of experimental variables. The equations required to carry out this calculation are derived in many books on rheology including [12, 13, 14, 16].

First, it can be shown that for a specific fluid at a fixed temperature, there is a unique relationship between the shear stress at the wall, τ_w, and the apparent wall shear rate. This means that if we obtain pressure drop data for a variety of flow rates, tube lengths and tube radii, they should all fall on a single curve when a plot of $[4Q/\pi R^3]$ versus $[(-\Delta P)R/2L]$ is prepared. A logarithmic plot is often used, and it can further be shown that the shear rate at the wall is related to the slope of the curve on such a plot. Equation 4-19,

$$\dot{\gamma}_w = \frac{4Q}{\pi R^3} \left(\frac{3 + b}{4} \right) \tag{4-19a}$$

where

$$b \equiv \frac{d \ln \left(\dfrac{4Q}{\pi R^3} \right)}{d \ln \left(\dfrac{-\Delta P \cdot R}{2L} \right)} = \frac{d \log \left(\dfrac{4Q}{\pi R^3} \right)}{d \log \left(\dfrac{-\Delta P \cdot R}{2L} \right)} \tag{4-19b}$$

has been variously associated with the names of Weissenberg, Rabinowitsch [51] and Mooney, while the bracketed term involving b is usually called the "Rabinowitsch correction." By comparison with Equation 4-13, we see that this term represents the deviation from Newtonian behavior. For a power law fluid, it is obvious from Equation 4-18 that $b = 1/n$.

From the above discussion, we see that a plot of the logarithm of the apparent wall shear rate versus the logarithm of the wall shear stress reveals, at a glance, the general nature of the viscosity function. If the data fall on a straight line with a slope of 1, Newtonian behavior is indicated. If they fall on a straight line with a slope not equal to 1, then power law behavior is indicated, and the slope is equal to $1/n$. Curvature implies that the behavior is neither Newtonian nor power law, and Equation 4-19 must be used to determine the true wall shear rate.

Once the shear rate has been determined by use of Equation 4-19, the viscosity can be calculated by use of Equations 4-10 and 4-14. Thus, these equations provide a basis for the determination of the viscosity of a non-Newtonian fluid on the basis of tube flow data. However, this requires differentiation of the data so that it is not possible to calculate a value of viscosity using data from a single experiment.

The procedure outlined above for determining the viscosity is applicable to the case of molten polymers. However, since these materials are elastic, there is an ambiguity in the use of the term pressure and the symbol ΔP. Thus, wherever the symbol P appears in the above equations, it should be replaced by the average axial normal stress over the tube cross-sectional area, A.

$$P \rightarrow - \int_A \frac{\sigma_{zz} dA}{A} \equiv - \bar{\sigma}_{zz} \tag{4-20}$$

$$\Delta P \rightarrow - \Delta \bar{\sigma}_{zz} \tag{4-21}$$

If stress measurements are made in terms of the force exerted by a piston or the pressure in a low viscosity fluid which has an interface with the melt, this normal stress can be easily related to the measured quantities, although, as is shown in the next section, end effects must be accounted for.

If measurements of wall pressure drop, ΔP_w, are made, then a relationship between this quantity and the average axial normal stress is needed. We note that for fully developed flow, the variation of stress with radial position and the values of the normal stress differences are independent of z. This means that

$$\Delta \bar{\sigma}_{zz} = \Delta \sigma_{zz}(r = R) = \Delta \sigma_{rr}(r = R). \tag{4-22}$$

But $\sigma_{rr}(r = R)$ is just the negative of the wall pressure, so that

$$\Delta \bar{\sigma}_{zz} = -\Delta P_w. \tag{4-23}$$

Now we can rewrite Equation 4-10 in a form appropriate for elastic fluids.

$$\tau_w = \frac{(-\Delta P_w) R}{2L} \tag{4-24}$$

If the wall pressure is measured by means of a transducer communicating with the tube wall by means of a small hole, then the measurement will be subject to pressure hole error. However, for fully developed flow, the pressure hole error is independent of z so that

$$\Delta P_w = \Delta P_m. \tag{4-25}$$

4.3.2 Entrance Effects and End Corrections

As we have seen, the shear stress can be determined by measuring the wall pressure at two axial positions, both in the fully developed flow region. However, a more common procedure is to measure the driving pressure, P_d, in the barrel and to assume that the pressure at the outlet of the capillary is equal to the ambient pressure, P_a. If the pressure is monitored by means of a load cell that measures the load on a driving piston (plunger), then the measured quantity is the driving force, F_d, and this is related to the driving pressure as follows, if friction between the plunger and the barrel is neglected.

$$P_d = \frac{F_d}{\pi R_b^2} \tag{4-26}$$

Thus, the pressure drop $(-\Delta P_w)$ in Equation 4-24 is replaced by the pressure difference $(P_d - P_a)$ or, since for melts P_d is nearly always much larger than P_a, $(-\Delta P_w)$ is simply replaced by P_d. However, this is clearly not the wall pressure drop that one would observe for fully developed flow in a length of capillary equal to L. It is, therefore, necessary to make some kind of "end correction" when using P_d to calculate τ_w.

First, let us examine the various reasons why the driving pressure is not equal to the wall pressure drop that would exist in fully developed flow through a length of capillary equal to L.

1. Some polymer may stick to the wall of the barrel and be sheared between the wall and the piston driving the flow. In addition, the piston itself will rub against the barrel wall unless it is perfectly straight, properly aligned and of the correct size. To provide a good seal, carefully sized piston tips are used, but if these respond to elevated temperatures in a different way than the barrel, then dimensional tolerances will be altered when the operating temperature is changed. This can result in either the scoring of the barrel wall or the opening up of a gap for the polymer to flow into. The presence of piston friction can be detected by operating the rheometer at the testing temperature without any polymer in the barrel. This source of error is especially important at low shear rates or for low viscosity materials, because, in these cases, the wall shear stress is small.

2. The increase in kinetic energy as the fluid moves from the barrel to the capillary will result in a drop in pressure. While this effect can be significant in the case of low viscosity liquids, it is likely to be relatively unimportant in the case of high viscosity melts.

3. The flow of polymer through the barrel will have associated with it a wall shear stress and a corresponding pressure drop. However, since the barrel has a substantially larger diameter than the capillary, this pressure drop is usually small compared to that resulting from the wall shear stress in the capillary. If it is not negligible, it will result in a small decrease in driving pressure during an experiment conducted at constant flow rate, and Metzger and Knox [52] have suggested a procedure for taking this into account.

4. As fluid approaches the entrance to the capillary, it starts to undergo a change in velocity distribution, reaching a practically fully developed profile only after moving some distance down the capillary. In the neighborhood of the entrance, the wall shear stress is larger than for the case of fully developed flow, and this gives rise to a wall pressure gradient larger than that corresponding to fully developed flow. This phenomenon is observed in the case of inelastic liquids, but it leads to especially large pressure drops in the case of elastic liquids. The excess pressure drop resulting from velocity rearrangements at the entrance is called the

"entrance pressure drop," while the length of capillary required for the velocity profile to approach its fully developed form is called the "entrance length." The entrance pressure drop generally makes a substantial contribution to the driving pressure, and it cannot be ignored in the analysis of capillary flow data.

5. As the fluid approaches the exit of the capillary, its velocity profile may change in anticipation of the disappearance of the confining wall. This effect is expected to be most prominent at low Reynolds numbers typical of the flow of molten polymers. The wall pressure gradient in this region would be larger than that for fully developed flow, and this would make an additional contribution to the total pressure drop beyond that which would be observed for fully developed flow in a capillary of length L.

The type of wall pressure distribution actually observed for capillary flow of molten polymers is sketched in Figure 4-7. Plots of actual exper-

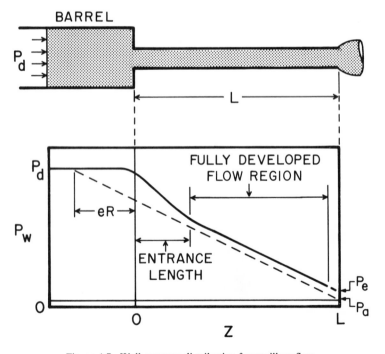

Figure 4-7. Wall pressure distribution for capillary flow.

imental data can be found in the book by Han [6]. It is not possible to measure wall pressures right up to the exit, and there is thus some uncertainty as to exactly what happens in the exit region.

The phenomena described under 1, 2, and 3 above are usually neglected in the treatment of capillary flow data. Items 4 and 5 involve the effects of velocity redistribution at the ends of the capillary, and one way of accounting for these is to measure the pressure drop for two capillaries of different lengths, at the same flow rate. Then the pressure gradient for fully developed flow (FDF) can be calculated as follows.

$$\frac{2\tau_w}{R} = \left(\frac{-\Delta P_w}{L}\right)_{FDF} = \frac{P_{d,2} - P_{d,1}}{L_2 - L_1} \qquad (4\text{-}27)$$

This procedure requires that both capillaries be of sufficient length to ensure that there is in each a length over which the flow is fully developed. To be sure that this is the case, a third capillary, longer than the first two, can be used, again at the same flow rate. The calculated wall pressure gradient for fully developed flow should be the same no matter which combination of capillaries is used.

An alternative procedure involves the use of an end correction, e, defined by Equation 4-28.

$$\tau_w = \frac{P_d}{2(L/R + e)} \qquad (4\text{-}28)$$

The product of e and R is the length of fully developed capillary flow having a pressure drop equal to the excess pressure drop resulting from end effects. Its significance is illustrated in Figure 4-7. (The product, eR, is not to be confused with the entrance length.) Bagley [53] applied this technique to polymer melts. He used a variety of capillaries having a wide range of L/R values and plotted P_d versus L/R with the wall shear rate as a parameter. Straight lines were obtained, implying that the end correction is independent of L/R. The end correction appropriate for a given shear rate was then determined by extrapolating one of these lines to zero driving pressure and noting the intercept on the abscissa.

Bagley found that the P_d versus $\dot{\gamma}_w$ curves continued to be straight even at quite small values of L/R. This is surprising and implies that the entrance length is very small. However, as a general rule, one should expect some curvature at small values of L/R, and the extrapolation should be made from the straight portion of the curve.

Curvature of the lines on a "Bagley plot" at large values of L/R, especially at high shear rates, may indicate a viscous heating problem. Such curvature can also result from the pressure dependence of viscosity [54, 55].

Bagley [53] assumed the velocity profile to be that of a Newtonian fluid and used Equation 4-13 to calculate the wall shear rate. This assumption is not generally valid for polymeric liquids. Thus, the Bagley plot should be prepared by plotting P_d versus L/R with the apparent shear rate, $(4Q/\pi R^3)$, as a parameter. The appearance of such a graph is sketched in Figure 4-8.

Once the end correction, e, has been found for each apparent shear rate, the actual shear rate at the wall can be found by the use of Equation 4-19. A form of Equation 4-19b, written in terms of the quantities defined in this section, is given below.

$$b = \frac{d \log \left(\dfrac{4Q}{\pi R^3} \right)}{d \log \left[\dfrac{P_d}{2 \left(\dfrac{L}{R} + e \right)} \right]} \qquad (4\text{-}29)$$

Because rotational rheometers suitable for the study of time and frequency dependent behavior are limited to use at low shear stresses, it would be useful to be able to interpret capillary flow data in such a way that elastic properties could be determined. Philipoff and Gaskins [56] have suggested that the entrance correction, e, could be considered to be made up of two components, one of which, n, is a viscous effect and the other is an elastic effect that can be related to the recoverable shear, S_r, defined in Equation 2-54.

$$e = n + S_r/2 \qquad (4\text{-}30)$$

Bagley [57] combined this result with "Hooke's law in shear" (see Equation 2-55) to derive a linear relationship between the end correction and the wall shear stress.

However, there is a strong convergence of streamlines at the entrance

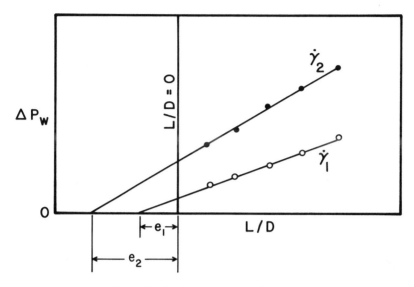

Figure 4-8. Bagley plots for two shear rates.

to a capillary, so there is a high degree of stretching along streamlines, and the flow there is quite different from simple shear. Furthermore, Hooke's law in shear cannot be relied on at high shear rates where linear viscoelasticity is not observed. Thus, one cannot expect these simple relationships between the end correction and recoverable shear or wall shear stress to have general validity.

In addition, it has been noted that the extrapolation of wall shear stress versus z data to $z = L$ yields a non-zero intercept for molten polymers, and Han and Charles [58] have proposed a modification of Equation 4-28 to account for the exit effect. On the other hand, Han [59] has demonstrated that if the capillary is sufficiently long that both the entrance and exit effects are independent of L/D, then the use of Equation 4-28 is justified. But this means that the end correction e includes both entrance and exit corrections, and this casts further doubt on the validity of relationships between e and recoverable shear.

4.3.3 Viscous Heating

Rheological properties depend on temperature and pressure, and variations in either of these complicate the interpretation of experimental data. As was pointed out in Chapter 3, viscous heating is inevitable in the deformation of viscous fluids, and attempts to maintain constant wall temperatures will always result in temperature gradients.* Thus, we must design experiments to keep the temperature nonuniformity within acceptable bounds.

It is useful to be able to estimate the extent of the temperature non-uniformity resulting from viscous heating. Bird [61], Martin [62] and Cox and Macosko [63] have calculated temperature profiles for power law fluids. Galili *et al.* [64] and Hieber [65] included the effect of pressure-dependent viscosity.

To make these calculations, it is necessary to make an assumption about what happens at the wall of the capillary, and two limiting cases are usually considered. In the "isothermal case," the wall is assumed to be at a uniform temperature, T_0, and, in the "adiabatic case," it is assumed that there is no heat transfer at the wall (i.e., that the wall is perfectly insulated). Figure 4-9 shows sketches of the types of profiles obtained for the isothermal case, at the entrance where the fluid is assumed to have a uniform temperature (curve *A*), at an intermediate distance downstream (curve *B*) and at a distance sufficient for a fully developed temperature distribution to be established (curve *C*).

Figure 4-10 shows the type of temperature profile obtained for the adiabatic case. In this case, there is no "fully developed" profile, as heat

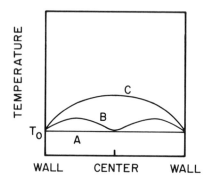
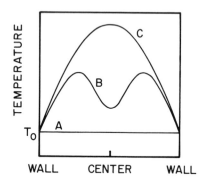

Figure 4-9. Temperature distributions for capillary flow with isothermal wall: Curve *A*, at the entrance; Curve *B*, some distance from the entrance; Curve *C*, far from the entrance. (a) Low flow rate. (b) High flow rate.

---

*Viscous heating is itself sometimes used as a basis for the study of melt flow. For example, Daryanani *et al.* [60] measured frictional heat in a capillary rheometer.

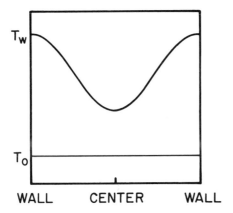

Figure 4-10. Temperature distribution for adiabatic flow in a capillary. T_0 is the entrance temperature and T_w is the wall temperature some distance downstream.

continues to accumulate indefinitely as the fluid passes along the tube, and the wall temperature rises continuously.

Cox and Macosko [63] and Gerrard *et al.* [66] have noted, on the basis of experimental studies, that it is more difficult to maintain a uniform wall temperature than has generally been assumed. In addition, the fully developed profile is not of practical interest, because it is only achieved far from the entrance in very long capillaries. Therefore, the actual temperature distribution is probably intermediate between those shown in Figures 4-9 and 4-10.

The prudent rheologist should make some effort to estimate the extent of the temperature variation due to viscous heating for each experiment that is carried out. However, the use of the general equations resulting from the analyses cited above is a tedious procedure. A rough estimate of the maximum temperature rise due to viscous heating at a particular distance z from the entrance can be made by the use of the nomograph shown in Figure 4-11. This estimating aid was developed by Middleman [67] on the basis of the calculations of Bird [61] for the adiabatic flow of a power law fluid. Middleman suggests that the nomograph will yield an estimate of the temperature rise at the wall within a factor of 2 of the exact value predicted by the complete equations.

To use the nomograph, start by locating the correct points on the wall shear stress and apparent wall shear rate scales at the left. Set a straightedge on both these points and note the intersection with index

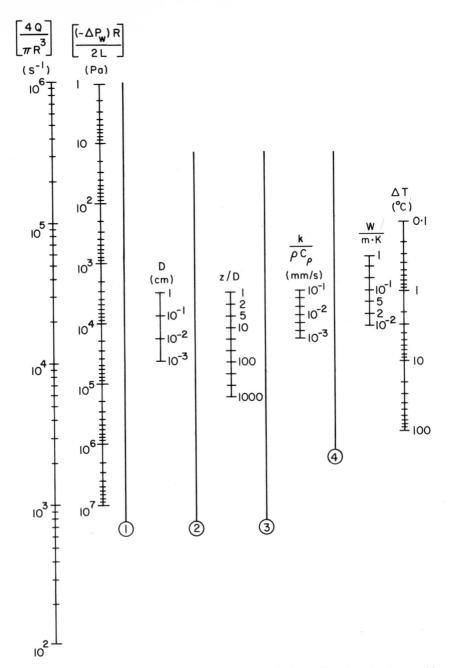

Figure 4-11. Nomograph for estimation of temperature rise in capillary flow. (Redrawn with minor revision from S. Middleman, *The Flow of High Polymers*. Interscience, NY (1968). Used with the permission of John Wiley & Sons.)

line number 1. Use this intersection as a pivot and reset the straightedge to pass through this point as well as the appropriate point on the D scale. This will define an intersection on index line number 2, which becomes the next pivot point, and so on from left to right. The resulting ΔT is the wall temperature less the temperature at $z = 0$ at a distance z from the entrance. Table 4-1 lists some typical values for the melt properties required to use the nomograph. They are taken from a tabulation originally published by Middleman [67].

Winter [68] has presented generalized plots of the temperature increase for tubes, slits and annuli, which are calculated for a power index of 0.4, taking into account the temperature dependence of viscosity. These will usually provide a more precise estimate of the temperature rise than is given by Figure 4-11.

It is obvious that the problems arising from viscous heating are minimized by the use of a capillary which is short and small in diameter. However, the use of a short capillary exacerbates the problems arising from entrance effects, so that some compromise must be made.

4.3.4 Pressure Effects

Pressure effects are especially troublesome in the use of capillary rheometers because the flow is pressure driven, and pressure gradients are unavoidable. As the driving pressure is increased to allow the study of higher shear rates, a point will always be reached where the variation of viscosity with pressure cannot be neglected. Penwell and Porter [69], for example, found that the viscosity of polystyrene seems to increase

Table 4-1. Typical Values of Thermal Properties for Molten Polymers*

POLYMER SYSTEM	THERMAL CONDUCTIVITY, k(W/m·K)	$k/\rho C_p$ (m²/s)
Polybutadiene/styrene copolymer	0.21	0.12
Polybutadiene	0.12	0.074
Polyethylene	0.33	0.13
Polyethylene terephthalate	0.12	0.051
Polystyrene	0.12	0.062

*Data taken from Middleman [67].

with shear rate when studied in a capillary rheometer over a certain range of temperature and shear rate, and they were able to relate this apparent increase to the effect of pressure on the glass transition temperature.

There are two possible approaches to this problem, the simplest being to limit experimental conditions to situations in which the pressure effect is thought to be negligible. The easiest way to judge this is to see if the Bagley plot is linear. Alternatively, one can analyze the experimental data in such a way as to take into account the pressure dependence of viscosity, and Penwell et al. [54] have done this by assuming a form for the relationship between pressure and viscosity.

Kamal and Nyun [55] have proposed a procedure for treating data that allows for end effects, viscous heating and pressure dependence. This procedure involves the differentiation of data with respect to both length and wall temperature (assumed constant) to determine the wall shear stress corresponding to a particular melt temperature and pressure. In addition, the differentiation indicated in Equation 4-19 must be carried out to determine the true wall shear rate. Thus, data for a variety of wall temperatures, capillary lengths and flow rates are required in order to use the procedure, and it is useful only when a rather extensive characterization is desired. However, it has the advantages of requiring no constitutive assumptions and yielding the pressure dependence of the viscosity.

Hieber [65] has raised a question with regard to the way in which viscous heating is accounted for in the procedure outlined above, and this would appear to stem from the authors' assumption that there is an initial length of capillary over which the flow is isothermal. However, for long capillaries this probably does not affect the results significantly.

4.3.5 Flow Instabilities

The equations given in Section 4.3.1 for calculating shear rate and viscosity from capillary flow data are based on the assumptions that the flow is steady and that the velocity is a smooth function of radius, going to zero at the wall (the "no slip" condition). However, under certain conditions, the flow of melt in a capillary can become unsteady, leading to a severe distortion in the shape of the extrudate. This tends to occur at high values of shear stress, so that increases in shear rate or viscosity or a decrease in temperature will increase the likelihood of its occurrence. This phenomenon, sometimes called "melt fracture," has been studied extensively, because it poses a severe restriction not only on the use of capillary rheometers but on certain industrial extrusion processes as well. The papers by Bergem [70] and Oyanagi [71] contain interesting photographs and sketches of the detailed nature of extrudate distortions.

Another phenomenon that is sometimes observed in capillary flow of molten polymers, especially high density polyethylene, is the sudden increase of flow rate when the driving pressure reaches a certain level. This has been studied, for example, by Rudin and Chang [72]. This seems to be related to the slip of polymer at or near the wall, resulting from a loss of adhesion between the polymer and the wall, or of a loss of cohesion along a cylindrical surface within the polymer but very near the wall [73]. An unstable situation can develop in which the polymer flow rate alternates between higher and lower values, and this has been referred to as the "stick-slip" phenomenon. It results in a marked irregularity in the shape of the extrudate [74].

It has been sometimes observed [72] that at flow rates somewhat above those at which the stick-slip phenomenon occurs, a smooth extrudate is again produced. However, it is likely that this results from the permanent detachment of the melt from the wall.

It is unsteady flow at the inlet to the capillary that is thought to be the ultimate origin of most, if not all, extrudate distortion [75]. Therefore, its occurrence can usually be deferred by the use of a very smoothly shaped transition from the barrel to the capillary in place of the simpler "flat" entry usually used.

When extrudate distortion or slip flow is observed, capillary flow data cannot be used to calculate reliable values for rheological properties. However, the conditions under which the distortion occurs should be recorded, as this information may be of practical importance.

4.3.6 Polymer Degradation in Capillary Flow

It seems likely that high molecular weight polymers would undergo some mechanically-induced degradation at high shear rates. Arisawa and Porter [76] did indeed observe that the low shear rate viscosity and molecular weight decreased in successive passes of a high molecular weight polystyrene through a capillary viscometer at high shear rate and that the presence of oxygen accelerated the degradation process.

The high shear rate viscosity, however, is rather insensitive to the molecular weight, so this phenomenon does not appear to pose a serious problem for rheologists. Of course, if the shear rate is reduced, data should not be recorded until all the material previously subjected to a high shear rate has cleared the capillary.

If the polymers of interest undergo rapid thermal degradation at melt temperatures when exposed to moisture and oxygen, then special precautions must be taken to minimize the exposure of the melt to these gases. Wissbrun and Zahorchak [77] have suggested a technique for carrying out capillary rheometer tests that allows for the blanketing of the polymer with an inert gas.

4.3.7 Summary of Procedures for Treating Capillary Flow Data

A summary of the steps to be taken in interpreting the results of capillary flow measurements is given below.

1. Estimate the maximum temperature nonuniformity by the use of Figure 4-11 or the more accurate equations of Bird [61] or the dimensionless plots of Winter [68].
2. Watch for extrudate distortion, and record its occurrence.
3. Make a plot of the driving pressure, P_d, versus the apparent shear rate, $(4Q/\pi R^3)$, for various values of L/R.
4. Tabulate values of P_d corresponding to various values of L/R at a fixed value of $(4Q/\pi R^3)$, and prepare a Bagley plot. Determine the end correction, e, as a function of $(4Q/\pi R^3)$. If the Bagley plot is not linear over a significant range of (L/R), then the end correction cannot be determined, and the possible role of pressure-dependent viscosity should be examined.
5. Determine the Rabinowitsch correction factor, b, by use of Equation 4-29 and the shear rate at the wall by use of Equation 4-19a.
6. Calculate the shear stress using Equation 4-28.

Negami and Wargin [78] have shown how most, if not all, of the graphical analysis involved in steps 3 to 6 can be eliminated by the use of a digital computer procedure.

If, instead of the driving pressure, P_d, the actual wall pressure, P_w, is measured at two axial positions, then steps 3 and 4 are unnecessary, and the wall shear stress can be calculated directly from Equation 4-24.

4.4 CAPILLARY RHEOMETERS

The capillary rheometer is the single most popular type of apparatus used in melt rheometry, and many designs have been described in the literature. Some of these having unique design features will be described in this section.

The primary use of capillary rheometers is the measurement of viscosity at high shear rates, although mention should be made here of the "rising column" viscometer designed by Elliott [79] to determine the viscosity of melts or concentrated solutions at very low shear rates. The polymer is contained in a glass test tube into which is inserted a glass capillary with a known inside diameter and with several horizontal marks scribed on it. A vacuum is applied to the upper end of the capillary, and the time required for the melt to rise from one mark to another in the capillary is measured. Elliott shows how this device can also be used to measure density and surface tension.

The rising column viscometer employs atmospheric pressure as the driving force for flow. Other driving mechanisms used for capillary rheometers include gravity acting on a suspended body, compressed gas, hydraulic systems and electromechanical drives. In the weight-driven devices, the maximum driving pressure is limited by the inconvenience of handling very large metal weights. The use of compressed gas, acting either directly or through a pressure-multiplying pneumatic cylinder, allows the continuous variation of driving pressure up to large values without the use of weights. Servohydraulic and electromechanical drives allow for increased flexibility, improved control of variables and operation at prescribed shear rates.

The unique features of a number of capillary rheometers that have been described in the research literature are set out in the following three subsections.

4.4.1 Weight-Driven Capillary Flow Testers

Because they are relatively inexpensive and simple to operate, weight-driven capillary flow testers are the devices most widely used in industry for the characterization of molten polymers. The "flow indexer" and the Rossi-Peakes flow tester will be described in this section. There are standard industrial tests based on both these devices, and commercial versions of them are described in Sections 9.2 and 9.3.

Capillary rheometers employing a weighted piston to provide the driving pressure have been used for many years (for example, by Hunter and Oakes [80] and by Tordella and Jolly [81]). Wiley [82] designed a miniaturized version for use with samples as small as 0.2 grams.

Based primarily on the work of Tordella and others at the Dupont Company, the American Society for Testing and Materials (ASTM) issued a standard test method based on the use of such a device. The essential features of the device described in ASTM standard D1238* are shown in Figure 4.12. The polymer (6) is contained in a barrel equipped with a thermometer (3) and surrounded by an electrical heater (4) and an insulating jacket (5). A weight (1) drives a plunger (2), which forces the melt through the die (7).

The ASTM standard prescribes the exact geometry of the cylinder (barrel) and die, the latter having an L/D of about 4. A variety of temperatures and total loads (including piston) are specified for use with different materials. Condition "E" is the one most commonly used with polyethylene and involves a temperature of 190°C and a total load of 2.16 kilograms. The standard procedure involves the determination of the amount of polymer extruded in 10 minutes. The flow rate, expressed as grams/10 minutes (equivalent to dg/minute), is the result reported. When condition "E" is used to characterize polyethylene, this quantity is commonly called the "melt index," and devices designed according to ASTM D1238 are often called "melt indexers." Obviously, the flow rate or "index" is higher for low viscosity materials.

Because of their simplicity and relatively low cost, testers designed in accordance with ASTM D1238 are widely used for quality control and for distinguishing between members of a single family of polymers. However, without extensive modification, these testers cannot give reliable values of the viscosity, and the ASTM Standard contains this statement: "The flow rate obtained with the extrusion plastometer is not a fundamental polymer property. It is an empirically defined parameter critically influenced by the physical properties and molecular structure of the polymer and the conditions of measurement." The reference to conditions of measurement is reinforced by a warning: "Relatively minor changes in the design and arrangements of the component parts have been shown to cause differences in results between laboratories."

---

*The corresponding standards for a number of other countries are listed in Chapter 9.

First of all, there is likely to be a significant entrance effect, and this cannot be determined unless several capillaries, having different values of *L/D*, are used. In any event, an *L/D* of 4 is probably too short for a fully developed flow to be achieved. Kowalski [83] recommends the substitution of a die having an *L/D* of 16 or more to "decrease the influence of entrance and frictional losses." Furthermore, as noted by Schreiber and Rudin [84], this type of capillary extruder is subject to

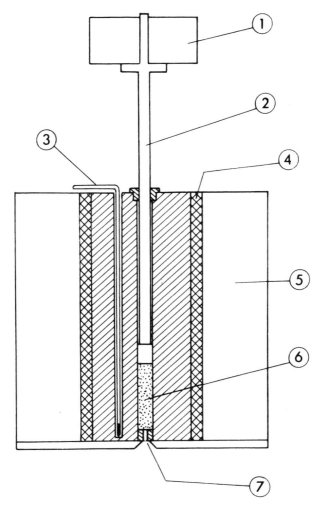

Figure 4-12. Essential elements of extrusion plastometer described in ASTM D1238. (Reproduced from J. L. Leblanc [16] with permission of Editions CEBEDOC, Liege.)

variation in flow rate during the course of an experiment. Thus, the flow rate obtained cannot be interpreted in terms of well-defined rheological properties.

In summary, the extrusion plastometer or "melt indexer" is a useful tool for comparing resins of the same type. However, in comparing resins not of the same family, it is important to note that two materials can have similar values of flow index and still have viscosity curves of quite different shape. This is illustrated in Figure 4-13.

It is essential that care be taken to perform the test exactly the same way each time. Since the measured parameter is defined only in terms of the details of the instrument design and experimental procedure, the only method of calibration is to compare observations made using a number of different, but supposedly identical, devices. If a quantity of resin having a known flow index is available, a melt indexer can be periodically checked for accuracy.

Another capillary flow device used to characterize molten polymers, particularly molding compounds, in industry is the Rossi-Peakes flow tester,* on which is based ASTM standard test method D569-59**: "Measuring the flow properties of thermoplastic molding compounds."

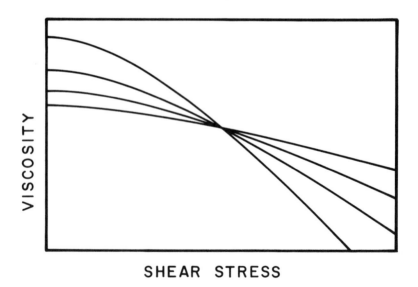

SHEAR STRESS

Figure 4-13. Viscosity curves for several materials having the same melt index.

*Covered by U.S. Patent 2.066.016.
**The corresponding British standard is B.S. 2782 (Method 105A).

A premolded, cylindrical specimen is placed at room temperature in a "charge chamber" (barrel) at the bottom of a steam-heated block which also contains a capillary with a diameter of 0.125 inch and a length of 1.5 inches. A piston is immediately raised into position and presses against the polymer, causing it to flow upward into the capillary. A specified upward driving pressure is supplied by means of weights suspended from pulleys. The rate of rise of polymer in the capillary is monitored by means of a follower rod which rests on the surface of the melt. The quantity reported is the temperature at which the flow in the first 2 minutes is 1 inch.

If polymer were preheated and allowed to fill the capillary, the Rossi-Peakes flow tester could be used as a crude capillary viscometer. In this case, it would be subject to most of the same sources of inaccuracy as the melt indexer, although the standard capillary for the Rossi-Peakes test has a much larger L/D. However, when used as specified in ASTM D569-59, this device yields a reading which depends not only on an unknown combination of rheological properties, but also on its melting and heat transfer characteristics. It is not intended to measure a material property, but to simulate, in some sense, the molding process.

A commercial version of the Rossi-Peakes tester is described in Section 9.3.

Maxwell and Jung [85] designed a weight-driven capillary rheometer for the study of the effect of pressure on melt viscosity. Like the Westover instrument described in the next section, it has barrel sections on both sides of the capillary, and both are fitted with pistons. A hydraulic hand pump, acting on the upper piston, supplies the elevated static pressure, while the driving pressure is applied to the lower piston by means of dead weight loading. This rheometer is simple, compact and portable.

4.4.2 Use of Compressed Gas to Drive Capillary Flow

Weight-driven capillary rheometers are, for practical reasons, limited to moderate driving pressures and thus to rather low shear rates. They therefore supply information only about the lower end of the viscosity curve and are of little value in distinguishing between resins whose viscosity curves are different in shape. This was recognized by Bagley and others working in the CIL laboratories,[*] and Bagley [53] has described the rheometer they developed as a result. In this device, which has come

---

*Canadian Industries Limited, Montreal.

to be called the "CIL high shear viscometer," the driving pressure is applied by means of pressurized nitrogen. Figure 4-14 shows the basic features of the instrument. The polymer (2) is heated by an electrical heater (4), and its temperature is monitored by two thermocouples (5). Nitrogen pressure is applied through a hose (7) and the melt flows through the die (1). To ensure an even flow of melt in the barrel, a steel ball (3), slightly smaller than the barrel, is used to transmit the gas pressure to the melt.

This instrument has been used in a number of important studies of

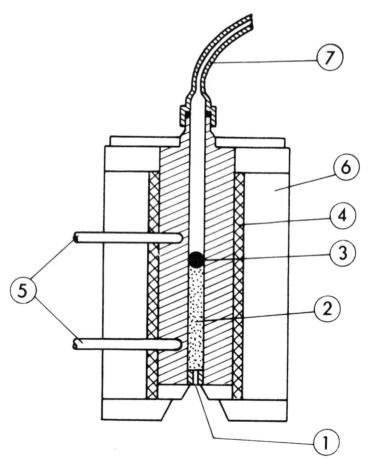

Figure 4-14. Essential elements of Bagley's rheometer [53]. (Reproduced from J. L. Leblanc [16] with the permission of Editions CEBEDOC, Liege.)

polymer melt flow and also as a quality control tester. A widely used standard test involves the use of the following die geometry and operating conditions.

Die diameter = 0.01925 inch
Die length   = 0.1760 inch
Pressure   = 1,500 psia
Temperature = 190°C

The flow rate is determined by weighing the polymer extruded in a fixed length of time and, for these particular conditions, the extrusion rate in grams/10 minutes is called the "CIL flow index."

Benbow and Lamb [86] designed a miniature version of the CIL viscometer for use with very small samples.

It has been reported that it is sometimes difficult to achieve steady flow in the CIL-type instrument. Schreiber [87] measured the flow rate as a continuous function of time and found that periods of 30 to 60 minutes were often required to achieve a constant flow rate. In some cases, no steady state was obtained even after one hour. Furthermore, Schreiber and Rudin [84] found that a sudden increase in pressure during extrusion actually decreased the flow rate of low density polyethylene. These observations demonstrate that a constant gas pressure does not necessarily lead to a constant flow rate, so that the existence of a steady state must be verified by flow measurements over several time periods.

For viscosity measurement at high shear rates, the CIL type viscometer has been largely replaced by rheometers with servohydraulic and electromechanical drive systems, in spite of their much greater complexity and cost. However, three commercial versions that are still available are described in Chapter 9.

Ramsteiner [88] has developed a pressure-driven capillary device that allows one to monitor with high precision the flow rate as a function of time. Gas pressure in the barrel forces the polymer through the die into an oil bath, which is heated electrically. The volume of oil displaced by the molten extrudate is measured electronically. A viewing port in the oil bath allows for observation of the behavior of the extrudate so that this device is useful for the study of isothermal extrudate swell and flow instabilities.

Another method for transmitting gas pressure to the melt in the barrel involves the use of a pneumatically operated piston. Tordella [89] used this technique to study extrudate irregularities, and Sieglaff [90] designed a rheometer for use with PVC based on the use of an air cylinder.* In the latter device, the driving pressure was measured by means of a load cell,** and the piston velocity by means of a velocity transducer. Mendelson [91] has described a technique for using such an instrument to study concentrated polymer solutions in volatile solvents at high temperatures.

Westover [92] designed a capillary rheometer for the measurement of melt viscosity at pressures up to 2,000 bars.† Figure 4-15 shows its basic features. Both ends of the capillary (1) are inserted into identical barrels, each equipped with a hydraulically operated piston (3). The difference between the pressures applied to the two pistons (5, 6) is the driving pressure for the flow. The precision with which this driving pressure can be determined is limited by the friction associated with the seals in the hydraulic pistons and by the fact that it is the difference between two large numbers. This type of rheometer has been used by Westover [93] and by Ito et al. [94]. Karl [95] has used a similar instrument designed for use at 5,000 bars.

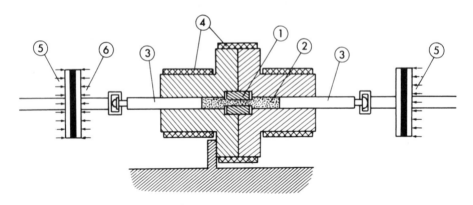

Figure 4-15. Westover's high pressure rheometer [92]. (Reproduced from J. L. Leblanc [16] with the permission of Editions CEBEDOC, Liege.)

---

*The Sieglaff-McKelvey rheometer is a commercial version of this rheometer. It is described in Chapter 9.
**These load cells were made by Lebow Associates (see Appendix B).
†A commercial rheometer based on the Westover design is described in Section 9.4.4.

4.4.3 Capillary Rheometers with Electric Motor Drive

If a piston activated by a mechanical drive running at a fixed speed is used to displace melt in the barrel, then a constant flow rate can be generated, and this is now a popular method for operating capillary rheometers designed for use with molten polymers.

In principle, any standard testing machine with a constant speed drive and a compression load cell can be used. Merz and Colwell [96] designed a rheometer assembly to fit into the frame of a standard Instron testing machine.* Its basic elements are shown in Figure 4-16.

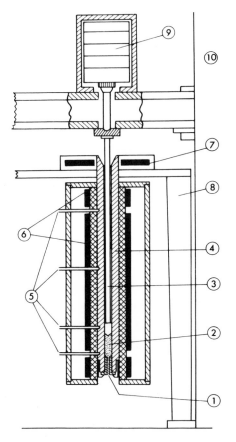

Figure 4-16. Merz-Colwell capillary rheometer [96]. (Reproduced from J. L. Leblanc [16] with permission of Editions CEBEDOC, Liege.)

---

*A commercial version of this device is described in Section 9.6.1.

The assembly is mounted on a support (8), which fits between the columns (10) of the testing machine. It consists of a barrel (4), electrical heaters (6) and thermocouples (5). The melt (2) is forced through the die (1) by a piston (3) connected to the load cell (9). Capillaries (dies) can be easily changed so that a wide range of L/D values can be employed, facilitating the determination of the end correction and allowing variations in wall shear rate between 1 and 12,000 s^{-1}.

Benbow [97] has described a simple capillary rheometer in which the piston is driven by an electric motor, and a commercial version of this instrument is described in Section 9.6.2. This device was modified by Hanson [98] to permit the exposure of the melt to shearing between two concentric cylinders prior to flowing into the capillary.

Kiselev et al. [99] used a constant flow rate capillary rheometer with several unique features. First, the capillary is able to slide freely in the barrel and is supported by means of a frame resting on a weight-measuring device. Thus, the force actually exerted on the capillary by the flowing polymer is measured directly. This eliminates piston and barrel losses, but an entrance correction is still necessary. Furthermore, a second length of capillary, also sliding freely in the barrel, is initially positioned some distance above the first where it does not contribute to the measured force. As polymer flows down the barrel, the second section of capillary eventually comes into contact with the first, effectively altering the total length of capillary and increasing the measured force accordingly. Thus, in a single run, two values of L can be employed, and if both are sufficient for fully developed flow to be achieved, Equation 4-27 can be used to calculate the wall shear stress.

Chalifoux and Meinecke [100] have described a capillary rheometer that has some features in common with Kiselev's instrument. However, in their rheometer, the capillary is in two sections with only the downstream section sliding freely. In this way, the entrance pressure drop makes no contribution to the force required to keep the sliding section from being displaced.

Mills [101] designed a miniature constant rate capillary rheometer for use when only a limited amount of resin is available, as capillary rheometers of standard design require about 20 grams of sample for each run. The driving pressure is measured by means of a pressure transducer mounted on the barrel wall.

Uhland [74] designed a capillary rheometer in which the barrel is charged by means of an extruder and the capillary is fitted with five wall-mounted pressure transducers.

An alternative to the electromechanical drive system is the use of a hydraulic drive, and several commercial rheometers using such a system are described in Section 9.5. A hydraulic drive lends itself readily to the use of a servo loop to control load, pressure or velocity. Load control is generally better for low shear rate work, while velocity control gives smoother flow at high shear rates.

4.5 SLIT FLOW

When fluid flows through a rectangular channel in which the width, w, is much larger than the thickness, h, the flow is two-dimensional over most of the cross-section, and the edges make a negligible contribution to the resistance to flow. The basic equations and entrance correction procedures are similar to those for capillary flow, but the difference in geometry has certain experimental advantages. Firstly, flush-mounted wall pressure transducers can be employed. This obviates the need for end corrections, which are expected to be greater for slit than for capillary flow, and makes possible the use of the techniques described in Sections 4.5.3 and 4.6 for the measurement of normal stress differences. Of course, since these transducers detect pressure through the deflection of a diaphragm, there will always be some irregularity in the wall where they are mounted. Secondly, the two-dimensional flow field facilitates the observation of flow in the channel and at its exit and the use of optical techniques such as birefringence.

Viscous heating and the resulting temperature nonuniformity is again a source of error as it is in all rheometers, although it is of special concern in devices designed for use at high shear rates. Langer and Werner [102] have developed a "correction diagram," based on a theory of Newtonian fluid flow, which provides a very rough estimate of the error arising from viscous heating in a slit rheometer. More rigorous analyses of viscous heating in slit flow have been reported by Ybarra and Eckert [103] and by Winter [68].

4.5.1 Basic Equations for Slit Rheometers

We will begin by giving the equations for the fully developed flow of a Newtonian fluid in a length L of channel having a width w and a thickness h, where w is much larger than h. A sketch of the slit flow geometry is shown in Figure 4-17.

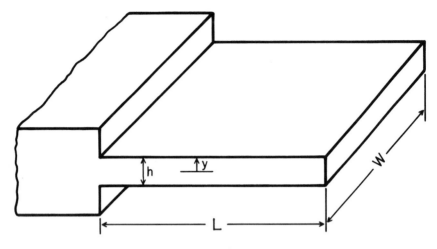

Figure 4-17. Slit flow geometry.

A force balance yields a relationship between the shear stress and the pressure gradient:

$$\tau_{yx} = \frac{\Delta P \cdot y}{L}$$

The magnitude of the shear stress at the wall, then, is given by

$$\tau_w \equiv -\tau_{yx}\left(y = \frac{h}{2}\right) = \left(\frac{-\Delta P}{L}\right)\frac{h}{2} \qquad (4\text{-}31)$$

If the shear stress at the wall were uniform all around the contour of the slit at a given value of x, the effect of the edge could easily be accounted for by multiplying the right-hand side of Equation 4-31 by the factor $w/(h + w)$. In fact, the shear stress is not uniform, but an average value can be defined as follows.

$$\bar{\tau}_w \equiv \frac{hw}{2(h + w)}\left(\frac{-\Delta P}{L}\right) \qquad (4\text{-}32)$$

For a Newtonian fluid, the edge effect can be calculated precisely, but for non-Newtonian fluids, no generalization is possible, and the usual procedure is to use a large aspect ratio, (w/h), together with Equation

4-31. Of course, knowledge of what constitutes a "large" aspect ratio must come from experience, and a value of 10 is usually considered sufficient.

The velocity profile for a Newtonian fluid is

$$v_x = \frac{3Q}{2hw}\left[1 - 4\left(\frac{y}{h}\right)^2\right]$$

(4-33)

and the magnitude of the shear rate at the wall is

$$\dot{\gamma}_w \equiv -\dot{\gamma}\left(y = \frac{h}{2}\right) = \frac{6Q}{h^2w}.$$

(4-34)

The viscosity of a Newtonian fluid can thus be calculated as follows.

$$\eta = \frac{\tau_w}{\dot{\gamma}_w} = \left(\frac{-\Delta P}{L}\right)\frac{h^3w}{12Q}$$

(4-35)

If the fluid is non-Newtonian and viscoelastic, the above equations are no longer valid. The stress equation can easily be adapted to the more general case by replacing P by P_w in accordance with the argument preceding Equation 4-24.

$$\tau_w = \frac{h(-\Delta P_w)}{2L}$$

(4-36)

The shear rate at the wall is no longer given by Equation 4-34, although this quantity is sometimes referred to as the "apparent" shear rate at the wall.

$$\dot{\gamma}_A \equiv \frac{6Q}{h^2w}$$

(4-37)

The actual shear rate can be determined by means of Equation 4-38a, a derivation of which has been given by Walters [14]:

$$\dot{\gamma}_w = \left(\frac{6Q}{h^2w}\right)\left(\frac{2 + \beta}{3}\right)$$

(4-38a)

where

$$\beta \equiv d \log \left(\frac{6Q}{wh^2} \right) \bigg/ d \ln \left[\frac{h}{2} \left(\frac{-\Delta P}{L} \right) \right] \qquad (4\text{-}38b)$$

As in the case of capillary rheometers, a plot of the logarithm of the apparent wall shear rate versus the logarithm of the wall shear stress reveals the general nature of the behavior of the viscosity function. If the data fall on a straight line with a slope of 1, Newtonian behavior is indicated. If they fall on a straight line with a slope not equal to 1, then power law behavior is indicated. Curvature implies that the behavior is neither Newtonian nor power law, and Equation 4-38 must be used to determine $\dot{\gamma}_w$.

If the fully developed velocity profile is maintained right up to the exit from the slit (i.e., if the flow remains "viscometric" at the exit), it is possible to show that the wall pressure in excess of the ambient pressure at the exit, P_{es}, is related to the first normal stress difference corresponding to the wall shear rate.*

$$N_1(\dot{\gamma}_w) = P_{es} + \tau_w \left(\frac{dP_{es}}{d\tau_w} \right) \qquad (4\text{-}39)$$

This relationship is discussed in some detail in Section 4.6.

The possibility of using the hole pressure error, P_H, to determine N_1 is considered in Section 4.5.3.

4.5.2 Slit Rheometers

Eswaran et al. [104] used a slit rheometer to determine the viscosity of a molten polymer, but measured the wall pressure at only two locations so that the linearity of the pressure distribution could not be verified. Wales et al. [105] added a third pressure transducer and also studied the effect of aspect ratio. Using slits with w/h equal to 10, 20 and 30, they found little effect, concluding that an aspect ratio of 10 is sufficient to justify the use of the slit equations. Han [106] reached this same conclusion by comparing measurements made with slits having ratios of 10 and 20. Han's rheometer consisted of a cylindrical block of alumi-

---

*For a derivation of this equation, see Han [6] or Walters [14].

num with the slot centered on its axis. Three holes in the block allowed for the insertion of pressure transducers. Han used an incorrect equation to calculate values of the first normal stress difference from extrapolated exit pressures, but later gave the correct equation [107].

Ehrmann [108] used an extruder to fill the barrel of his slit rheometer although a piston was used during operation of the rheometer. Langer and Werner [109] built a pressure-driven slit rheometer with an extra large reservoir. They calculated the effect of edge flow and used the result to estimate an edge flow correction for the apparent shear rate as calculated from Equation 4-37. Springer *et al.* [110] designed a twin-slit rheometer to be used, on-line, at the end of an extruder. The second slit allows for the balancing of the flow so that a variety of shear rates can be used without altering the extruder output rate. An incorrect equation was used to calculate the first normal stress difference.

Leblanc [111] used a slit die designed in such a way that the gap, h, could be easily changed by the use of spacers. The length was 85 millimeters, the width was 10 millimeters, and gaps of 0.5, 0.75, 1 and 1.5 millimeters were used. The driving pressure for the piston was supplied by means of a servohydraulic system, and the flow rate was determined by weighing extrudate. Combination temperature and pressure transducers* were inserted in the die at three axial locations. Leblanc [112, 113] also used this rheometer to study stress relaxation after cessation of shear. However, since the shear rate is not uniform and some bulk relaxation may occur, it is not clear how the $P_w(t)$ curve so obtained is related to the stress relaxation function. Robens and Winter [114] used a slit rheometer fed from an extruder, and Hansen and Jansma [115] used a similar arrangement to carry out tests of an engineering thermoplastic at temperatures up to 425°C in the absence of air.

4.5.3 Pressure Hole Error and Normal Stress Differences

In Section 3.2, it was explained that the pressure at the bottom of a hole in a wall over which a viscoelastic fluid is flowing, as shown in Figure 3-1, is not equal to the wall pressure. The pressure hole error, P_H, is defined as the difference between the measured pressure in the hole, P_m, and the correct wall pressure, P_w.

---

*These transducers were made by Dynisco (see Appendix B), who also supplied the transducers used by Han [106] and by Springer *et al.* [110]. These devices are especially well-suited for high temperature melt studies.

$$P_H \equiv P_m - P_w \qquad (4\text{-}40)$$

This quantity has been found to be negative for polymeric fluids, even when the hole is quite small.

Making a number of simplifying assumptions, Higashitani and Pritchard [116] derived a relationship between P_H and the first and second normal stress differences.

$$P_H = -\frac{1}{3} \int_0^{\tau_w} (N_1 - N_2) d\tau / \tau \qquad (4\text{-}41)$$

In particular, they assumed that the velocity and stress fields are symmetric about the center of the hole and that inertial effects can be neglected.

Equation 4-41 cannot be used directly to calculate the function $(N_1 - N_2)$ from pressure hole data, but it can be put in another form [117] which could, in principle, be so used.

$$(N_1 - N_2)_{\tau_w} = -3\tau_w \left(\frac{dP_H}{d\tau_w} \right) \qquad (4\text{-}42)$$

However, great precision and smoothness are required of data if they are to be differentiated to yield meaningful results, so that little use has as yet been made of Equation 4-42.

Baird [118] has pointed out that if the following assumptions are made,

$$N_2 \ll N_1$$
$$N_1 = a\tau^m$$

Equation 4-41 can be used to show that

$$P_H = -N_1/3m.$$

This implies that for a given material, the pressure hole error is proportional to the first normal stress difference. A general expression of this type of relationship is as follows.

$$P_H = -cN_1 \qquad (4\text{-}43)$$

Higashitani and Lodge [119] measured the pressure hole error by mounting pressure transducers opposite each other on the walls of a slit rheometer. One transducer was mounted flush with the wall and the other was mounted using a pressure hole. For several polymer solutions, they found that

$$P_H = -0.18N \pm 19\%.$$

Baird [118], using a similar device with some improvement, found that

$$P_H = -0.21N_1.$$

In general, we would expect that c would depend on molecular structure, molecular weight distribution, solution concentration and temperature. However, based on a review of data in the literature, Baird noted that for most of the polymer solutions that have been studied, c falls between 0.15 and 0.25 and thus does not appear to be strongly material-dependent.

Lodge [120, 121] proposed using a device of the sort described above as a "Stressmeter" for use as an in-line process instrument.* Lodge has also designed and constructed a Stressmeter for use with molten polymers [122]. He points out the importance of using a differential pressure measuring system. The quantity P_H is small in magnitude, and if P_m and P_w are determined from independent transducers and P_H obtained by subtraction, the result is subject to large errors. In his experiments with low density polyethylene melts, using this device, de Vargas [123] found good agreement with Equation 4-43. One problem that might arise in working with melts is the degradation of stagnant polymer in the pressure hole. Provision would then need to be made to flush out this material periodically. However, de Vargas found it unnecessary to bleed the hole when working with low density polyethylene at 150°C over a period of many hours.

Han [35] found P_H to be too small for reliable measurement for certain melts, although he did not use a differential pressure measuring system. Han and Yoo [124] studied the details of the flow of melts and solutions near a pressure hole and a slot and found significant deviations

---

*A commercial instrument based on this idea, the Seiscor/Lodge Stressmeter, has been developed for use with polymer solutions. (See Appendix B for the address of the Seiscor Division.)

from the flow assumed by Higashitani and Pritchard in deriving Equation 4-41. However, their experiments involved a small hole opposite a larger one, and the situation may be different in the case of a hole opposite a flat wall. Hou et al. [125] observed the flow of a melt over a slot and observed streamline curvature even at very low Reynolds numbers.

At the present time, it is not yet clear whether the measurement of pressure hole error will provide a useful basis for the rheological characterization of polymer melts.

4.6 EXIT PRESSURE AND NORMAL STRESS DIFFERENCES

It has been observed that the wall pressure at the exit of a capillary or slit is not equal to the ambient pressure. Han [6, 107] has proposed that this phenomenon be used as the basis for a method of measuring normal stress differences at shear rates higher than those accessible in rotational rheometers.* The use of "exit pressure" as a rheological material function has associated with it two important problems. First, there is the problem of how to measure the exit pressure and, second, there is the problem of relating exit pressure to viscometric functions.

The "exit pressure" is defined as the wall pressure at the exit of a capillary or slit.

$$P_{ec} \equiv [P_w(z = L)]_{\text{capillary}} \tag{4-44}$$
$$P_{es} \equiv [P_w(z = L)]_{\text{slit}} \tag{4-45}$$

It is not possible to measure the wall pressure right at the exit, so that exit pressure values are determined by measuring the wall pressure at several axial locations and then extrapolating these data to $z = L$. This extrapolation assumes that the pressure gradient remains linear right up to the exit, but at the low Reynolds numbers characteristic of melt flows, we expect, at least in the case of Newtonian fluids, that the effect of the exit will be felt some distance upstream. Nickell et al. [126] have calculated that, for a Newtonian fluid at very low Reynolds numbers, the velocity profile at the exit is significantly different from that for fully developed flow. Furthermore, Whipple and Hill [127] and Gottlieb and Bird [128] have observed the upstream velocity redistribution for the

---

*A commercial instrument based on this idea, the Seiscor/Han Rheometer, is described in Chapter 9.

capillary flow of polymer solutions. On the other hand, Han [6] has argued that this phenomenon is not important for molten polymers.

At the present time, this question is unresolved. On the one hand, it is clear that some rearrangement of the velocity profile occurs upstream from the exit so that viscometric flow does not obtain right up to the exit. On the other hand, the effect of this rearrangement on the wall pressure is not known, and it is conceivable that there may be, at least from an empirical point of view, some basis for the extrapolation of wall pressures to determine the exit pressure.

Another question arises in the measurement of wall pressures, particularly in capillaries, when it is not possible to mount the pressure transducer flush with the wall. If the transducer is mounted in a chamber which communicates with the flow channel by means of a hole in the wall, the measurements will be subject to pressure hole error, as described in Section 4.5.3. However, Han [35] has suggested, on the basis of some experimental evidence, that the pressure hole error is negligible for molten polymers. At the present time, there is no generally accepted answer to this question. Of course, if a flush-mounted transducer is used in slit flow, the question of pressure hole error does not arise, although the deflection of the diaphragm will always result in some small irregularity in the wall surface.

Now we turn to the problem of relating the exit pressure to viscometric functions. If the wall pressure varies linearly with axial position over some length of the capillary or slit, then the equations in Section 4.3 (for a capillary) or 4.5 (for a slit) can be used to calculate the wall shear stress and the viscosity. However, the reason for measuring exit pressure is to learn something about the material in addition to its viscosity. Han [6] has noted that exit pressure data have some similarity to equilibrium extrudate swell data, but extrudate swell involves a non-viscometric deformation and cannot be related to other rheological material functions unless rather specific constitutive assumptions are made. Therefore, it has been seen as desirable to relate exit pressure directly to normal stress differences, and this possibility is of special importance because there is at present no proven method for measuring normal stress differences at high shear rates.

Early attempts to derive relationships between exit flow characteristics and material functions contained errors, but the equations given by Han [6] and also by Walters [14] are now generally accepted as correct. If the flow is assumed to be viscometric right up to the exit, and if the inertia ("thrust") terms are neglected, on the basis of the high viscosity of molten polymers (and resulting low Reynolds number of the flow), then the following equations can be derived for capillary flow.

$$\left(N_1 + \frac{1}{2} N_2 \right)_w = P_{ec} + \frac{1}{2} \tau_w \left(\frac{\partial P_{ec}}{\partial \tau_w} \right) \tag{4-46}$$

$$(N_2)_w = \tau_w \frac{\partial}{\partial \tau_w} \left[P_{ec} - P_c(0, L) \right] \tag{4-47}$$

The quantity, $P_c(0, L)$, is the value at $r = 0$ and $z = L$ of a compressive, isotropic contribution, $P(r, z)$, to the total normal stress, σ_{rr}, for capillary flow.

$$\sigma_{rr}(r, z) = \tau_{rr}(r, z) - P(r, z) \tag{4-48}$$

Combining Equations 4-46 and 4-47 to eliminate the second normal stress difference, we obtain

$$(N_1)_w = P_{ec} + \frac{1}{2} \tau_w \frac{\partial P_c(0, L)}{\partial \tau_w}. \tag{4-49}$$

The quantity, $P_c(0, L)$, cannot be measured, so that Equation 4-49 cannot be used to determine $(N_1)_w$ from experimental exit pressure data. However, the combination $(N_1 + \frac{1}{2}N_2)_w$ can be determined by the use of capillary exit pressure and wall shear stress data.

Making the same assumptions as used above for a capillary, the following equation, shown first in Section 4.5.1, can be derived for slit flow.

$$(N_1)_w = P_{es} + \tau_w \left(\frac{\partial P_{es}}{\partial \tau_w} \right) \tag{4-39}$$

Thus, slit exit pressure data can be used to determine the first normal stress difference. If capillary exit pressure data at the same wall shear rates can be obtained, then once $(N_1)_w$ has been determined from slit data, Equation 4-46 can be used to calculate values of $(N_2)_w$.

On the basis of limited data, Han [107] has suggested that the capillary exit pressure is equal to the slit exit pressure at equivalent wall shear rates.

$$P_{es}(\dot\gamma_w) = P_{ec}(\dot\gamma_w) \qquad (4\text{-}50)$$

If this relationship is assumed to be true, then the following equations can be derived for capillary flow.

$$(N_1)_w = P_{ec} + \tau_w \left(\frac{\partial P_{ec}}{\partial \tau_w} \right) \qquad (4\text{-}51)$$

$$(N_2)_w = -\tau_w \left(\frac{\partial P_{ec}}{\partial \tau_w} \right) \qquad (4\text{-}52)$$

These results imply the following relationship between the exit pressure, P_{ec}, and the quantity, $P_c(0, L)$.

$$P_c(0, L) = 2P_{ec} \qquad (4\text{-}53)$$

Since $P_c(0, L)$ is not measurable, there is no way to verify this relationship experimentally except as an inference from the equality expressed by Equation 4-50, and currently there is no theoretical basis for it. However, if some confidence can be developed in Equation 4-50, then Equations 4-51 and 4-52 make it possible to determine both the first and second normal stress differences from capillary exit pressure data. Since the use of these equations involves differentiation, reliable results require smooth, precise data.

In summary, the use of exit pressure measurements to determine normal stress differences depends on the assumption that the flow is viscometric (parallel streamlines in this case) right up to the exit. This assumption is involved both in the extrapolation used to determine the exit pressure and in the equations used to relate exit pressure to normal stress differences. Furthermore, unless flush-mounted pressure transducers are employed (possible only with slit rheometers), the wall pressure measurements will be subject to hole pressure error. Finally, the calculation of both normal stress differences on the basis only of capillary exit pressure measurements requires the additional assumption that at equivalent wall shear rates the slit and capillary exit pressures are equal. However, it is possible to calculate the first normal stress difference from slit wall measurements, made using flush-mounted transducers, by making only the single assumption that the flow is viscometric at the exit.

Boger and Denn [129] have written a critical review of the use of exit pressure in rheometry, paying special attention to the possible effect of velocity rearrangement prior to the exit. They conclude that there is no way to establish, on general theoretical grounds, an upper limit on the error introduced by the velocity rearrangement effect. They suggest that the reliability of the exit pressure can only be established on the basis of empirical evidence. Based on their scrutiny of previously published exit pressure data, they conclude that such evidence does not exist at this time.

4.7 ANNULAR FLOW AND NORMAL STRESS DIFFERENCES

Tanner [130] has suggested several flows that may be useful in the determination of normal stress differences, and two of these involve flow in an annulus.

Axial flow in an annulus has been used by a number of researchers to obtain information about the second normal stress difference in polymer solutions. For example, Lobo and Osmers [131] used this technique, and they describe techniques for accounting for possible misalignment of the pressure holes and for pressure hole error. The basic rheometrical equation relates the difference between the wall pressures at the outer and inner walls to the second normal stress difference.

$$\Delta P_w \equiv (P_w^0 - P_w^i) = \int_{r_i}^{r_0} N_2 \frac{dr}{r} \qquad (4\text{-}54)$$

To carry out the integration, it is necessary to know the shear rate distribution, $\dot{\gamma}(r)$, and this requires the use of the viscosity function, $\eta(\dot{\gamma})$. Furthermore, it is necessary to hypothesize a form for the function $N_2(\tau)$. Thus, this technique can only be used to determine the material constants of an assumed function. Ehrmann [132] used an extruder to pump molten polymers into the annulus and assumed a modified power law for the $\Psi_2(\dot{\gamma})$ function to integrate Equation 4-54.

Okubo and Hori [133] also used axial annular flow to measure the second normal stress difference for a melt, assuming that the viscosity follows a power law. To avoid assuming a form for the relationship between N_2 and τ, however, they derived an approximate equation that allows the determination of N_2 at r_0 by differentiating the pressure dif-

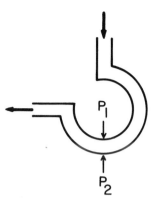

Figure 4-18. Geometry for circumferential flow in an annulus.

ference data. They note that the major challenge in the use of this experimental technique is the measurement of the small difference in wall pressures, ΔP_w.

For circumferential (azimuthal) flow around an annulus under the influence of a pressure gradient, the wall pressure difference, defined as above, is related to the first normal stress difference as follows.

$$\Delta P_w = - \int_{r_i}^{r_0} N_1 \frac{dr}{r} \tag{4-55}$$

It is again necessary to assume forms for the functions $\eta(\dot{\gamma})$ and $N_1(\dot{\gamma})$ in order to carry out the integration. Osmers and Lobo [134] developed an apparatus based on this type of flow for use with polymer solutions.* Geiger and Winter [135] used a similar technique to determine $N_1(\dot{\gamma})$ for molten polymers. Figure 4-18 is a sketch of the geometry they used. To detect the small pressure differences involved in this technique, Geiger [136] employed a "pressure difference transducer."

---

*The essential features of this apparatus are the subject of U.S. Patent 4027526.

Chapter 5
Rotational Flows Used in Melt Rheometry

Rotational flows involving a number of different geometries provide the basis for a number of rheometers that have proven very useful in the study of molten polymers. Most of these provide direct access to the shear surfaces, and they are thus easier to fill and clean than capillary rheometers. Furthermore, they can be used to measure a much wider spectrum of rheological properties. However, due to the centripetal acceleration that accompanies rotation, secondary flows are always present and become significant at sufficiently high speeds. Because of their very high viscosity, this is not a problem in the study of molten polymers. The limiting factor for melts is flow irregularity at the edge, which appears to be related to melt elasticity. The shear rate at which these irregularities become troublesome ranges from 0.1 to 10 s^{-1} depending on the exact geometry used and the material studied.

Other problems that arise in the use of rotational rheometers include instrument compliance and the friction associated with the rotation of the shaft supporting the rotating fixture. The latter problem is often solved by the use of air bearings, but this adds complexity and expense to the instrument and requires a reliable supply of clean dry air.

5.1 CONCENTRIC CYLINDER FLOW

The basic features of concentric cylinder geometry are shown in Figure 5-1. The fluid is contained in the gap between two cylinders sometimes referred to as the "bob" and the "cup." In a "Couette" type rheometer, the inner cylinder rotates, while in a "Searle" type, it is the outer cylinder that rotates.

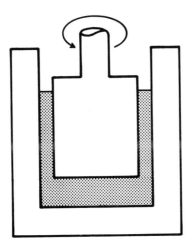

Figure 5-1. Concentric cylinder geometry.

5.1.1 Basic Equations

The equations presented here apply to either the Couette or Searle type instrument, with Ω representing the speed of rotation in either case. It is assumed that the streamlines are circles, which means that end effects and secondary flows are assumed to be absent. The validity of these assumptions is taken up in a later section.

The torque, M, required to turn the rotating cylinder, or to hold the stationary cylinder in place, is related to the shear stress at the wall of the inner cylinder, τ_i, as follows,

$$M(r_i) = 2\pi r_i^2 L \tau_i \qquad (5\text{-}1)$$

where r_i and L are the radius and length, respectively, of the inner cylinder.

While Equation 5-1 is valid for any fluid, an explicit relationship for the shear rate at the wall of the inner cylinder cannot be derived without assuming a form for the velocity function, $\eta(\dot{\gamma})$. Considering first the case of a Newtonian fluid, it can be shown that the shear rate at the inner wall, $\dot{\gamma}_i$, is as follows.

$$\dot{\gamma}_i = \frac{2\Omega}{[1 - (r_i/r_0)^2]} \qquad (5\text{-}2)$$

Thus, for a Newtonian fluid, the viscosity is given by the Margules equation.

$$\eta = \tau_i/\dot{\gamma}_i = \frac{M}{4\pi L\Omega}\left(\frac{1}{r_i^2} - \frac{1}{r_0^2}\right) \tag{5-3}$$

If the gap is very small compared to the radius of the cylinders, there is little variation in the shear rate across the gap, and an approximate, simplified equation is as follows.

$$\dot{\gamma} \approx \frac{\Omega r_i}{(r_0 - r_i)} \tag{5-4}$$

For example, if (r_i/r_0) is 0.99, the variation in shear rate across the gap is 2%. Thus, the flow generated when the gap spacing is small is a close approximation to simple shear.

Next, we consider the case of the power law fluid defined by Equation 2-10, for which the shear rate at the wall of the inner cylinder is as follows.

$$\dot{\gamma}_i = \frac{2\Omega}{n[1 - (r_i/r_0)^{2/n}]} \tag{5-5}$$

By setting $n = 1$, we obtain, of course, the Newtonian result, Equation 5-2. To determine n directly from experimental data, we can use an expression derivable from Equations 5-1 and 5-5 together with the definition of the viscosity.

$$n = \frac{d \log \tau_i}{d \log \Omega} \tag{5-6}$$

Thus, a useful procedure for the preliminary examination of data is to plot $\tau(r_i)$ versus Ω using logarithmic scales. If the data fall on a straight line, the fluid is following the power law, and if the slope is one, the behavior is Newtonian.

Finally, we turn to the case of general non-Newtonian behavior. In order to obtain an explicit equation for the shear rate without making any assumptions about the form of the viscosity function, it is necessary to write the shear rate as an infinite series. The trick here is to write the series in such a way that a good approximation is obtained by use of only a few terms. Yang and Krieger [137] have compared the various series that have been proposed, and they conclude that the most useful among these is one that has the leading terms given by Equation 5-7.

$$\dot{\gamma}(r_i) = \frac{2N\Omega}{[1 - (r_i/r_0)^{2N}]} \left[1 + \frac{f(x)dN}{N^2 \, d \, \ell n \, [\tau \, (r_1)]} + \cdots \right] \quad (5\text{-}7)$$

where

$$f(x) \equiv x(e^x - 1)^{-2} \, (xe^x - 2e^x + x + 2)/2$$

and

$$x \equiv -2N\ell n(r_i/r_0)$$

and

$$N \equiv \left[\frac{d \log \Omega}{d \log \tau_i} \right]$$

Note that, for a power law fluid $N = 1/n$ and Equation 5-7 reduces to Equation 5-5. Yang and Krieger [137] further recommend that, for most purposes, it is adequate to employ only the terms shown in Equation 5-7, although, for greater precision, they suggest adding one additional term.

However, rheometers for melts must have a small gap because of the viscous heating problem, and considering all the other sources of error, it seems superfluous to use anything more precise than the terms shown in Equation 5-7, and even the power law approximation will often be adequate. Moreover, since a small gap implies a small variation in shear rate, and this implies, in turn, a small variation in viscosity across the gap, Equation 5-4 may be a useful approximation where (r_i/r_0) is of the order of 0.99.

If a concentric cylinder rheometer is used to determine the viscoelastic properties, $\eta'(\omega)$ and $G'(\omega)$, either the inner or outer cylinder is made to oscillate with a sinusoidal angular displacement.

$$\phi = \phi_0 \sin(\omega t) \quad (5\text{-}8)$$

If the other cylinder is rigidly supported so that its angular displacement is essentially zero, then a measurement of the torque as a function of time, by means of an electronic torque transducer, provides the data necessary to calculate the desired properties. For example, if the outer cylinder is oscillated while the torque on the inner, stationary cylinder is measured, we can calculate the strain and stress at $r = r_i$ as follows, as long as ϕ_0 is small.

$$\gamma_i(t) = \frac{2\phi_0(t)}{[1 - (r_i/r_0)^2]} \qquad (5\text{-}9)$$

$$\tau_i(t) = M(t)/(2\pi r_i^2 L) \qquad (5\text{-}10)$$

The calculation of $\eta'(\omega)$ and $G'(\omega)$ procedes by use of the relationships given in Section 2.4.4.

Before reliable electronic torque transducers became available, the response of the inner cylinder was usually determined by suspending it from a torsion wire and following its angular displacement as a function of time. For this type of experiment, the desired material functions can be calculated by use of equations given, for example, by Walters [14].

The equations presented above involve only the shear stress, but it is also possible to relate the first normal stress difference to measurable quantities in concentric cylinder flow, as shown in Equation 5-11.

$$\Delta P_w \equiv P_w(r_0) - P_w(r_i) = \frac{1}{2} \int_{\tau_i}^{\tau_0} N_1(\tau) \frac{d\tau}{\tau} \qquad (5\text{-}11)$$

This equation, a derivation of which is given by Walters [14], can be inverted to give the following.*

$$N_1(\tau_0) - N_1(\tau)_i = \tau_i \frac{\partial(\Delta P_w)}{\partial \tau_i} \qquad (5\text{-}12)$$

Since we know that the first normal stress difference approaches zero as the shear stress decreases, this equation provides, in principle, a method for determining $N_1(\tau)$ and thus $N_1(\dot{\gamma})$, if $\eta(\dot{\gamma})$ is known. How-

---

*Inertia has been neglected in deriving this expression. The term that must be added when taking inertia into account is given by Walters ([14], p. 66).

ever, a large gap is needed to produce a measurable pressure difference, and the wall pressure measurements are likely to be subject to pressure hole error. While this technique has been employed in the study of polymer solutions, it has not been applied to melts.

5.1.2 Sources of Error for Concentric Cylinder Flow

Eccentricity. The equations given above relating shear rate to experimentally measured quantities are valid only when the inner and outer cylinders are concentric so that the flow is axially symmetrical. For Newtonian fluids, it can be shown theoretically that eccentricity increases the torque for given cylinder radii, speed and fluid viscosity, but there is no general theory for non-Newtonian or elastic liquids. Proper alignment is especially crucial when the gap spacing is small, as is likely to be the case in a rheometer designed for melts. Concentricity should be verified by direct measurement at operating temperature, if this is possible. In any event, periodic recalibration with a melt of known viscosity should reveal alignment problems.

End effects. The analysis of data usually assumes that there is no axial variation in the flow pattern, but at the ends of the inner cylinder this cannot be the case. For well-defined simple shapes, end effects can be calculated for Newtonian fluids, but this is not possible for elastic, non-Newtonian melts. The provision of a large space between the bottoms of the inner and outer cylinders has the effect of reducing the end effect for a Newtonian fluid, because a reduced shear rate is always associated with a proportionately reduced shear stress. However, the situation is not so simple for pseudoplastic liquids, because the viscosity is higher at lower shear rates.

One way of minimizing end effects is by the use of "guard rings," which are extensions above and below the section of cylinder on which the torque is measured. These guard rings move with this section but are not directly connected to it so that the end effects influence only the flow at the end of the guard rings. Kepes [138] has designed a concentric cylinder rheometer for melts that makes use of guard rings, but this technique has not been generally popular because it complicates the design of the rheometer and exacerbates cleaning problems.

Highgate and Whorlow [139] studied the end effect problem and concluded that there is no completely satisfactory way of eliminating or compensating for end effects for non-Newtonian fluids. However, if the gap ($r_0 - r_i$) is quite small, and the length of the inner cylinder not too small, it should be possible to limit the error due to end effects to about 2%.

Secondary flow. The calculation of rheological properties from concentric cylinder measurements is based on the assumption that the streamlines are circular. In the case of Newtonian fluids, it is known that secondary flows and turbulence can occur under certain circumstances. For example, if the inner cylinder is rotated, secondary flow will occur when the following inequality is satisfied.*

$$\left(\frac{\Omega r_i \rho h}{\eta} \right) \left(\frac{h}{r_i} \right)^{1/2} > 41 \tag{5-13}$$

When the outer cylinder is rotated, the flow is stable up to very high speeds where it eventually becomes turbulent.

The onset of secondary flow in the case of non-Newtonian and elastic liquids is not well understood, and one must simply be alert to the possibility of its occurrence as rotational speed is increased. The result is likely to be a discontinuity in the slope of the shear stress versus shear rate curve.

Rod climbing. When a cylinder is rotated while partially immersed in a viscoelastic liquid, the liquid tends to rise up at the surface of the cylinder.** This phenomenon, often called the "Weissenberg effect," can change the effective shearing area in a concentric cylinder rheometer, especially at high shear rates.

Viscous heating. If the gap spacing is small, the equations of Section 3.2 can be used to estimate the maximum temperature in the melt. For example, if the outer cylinder is maintained at a temperature T_w, and there is no heat transfer at the wall of the inner cylinder, the temperature rise can be estimated by using Equations 5-4 and 3-6, and the result is as follows.

$$T_{max} - T_w = \frac{\eta \Omega^2 r_i^2}{2k} \tag{5-14}$$

---

*This is an approximate form of the general criterion, valid only when the gap spacing is small.
**The possibility of using this phenomenon itself as a rheological tool is discussed in Chapter 7.

5.1.3 Some Concentric Cylinder Melt Rheometers

In this section, we will begin by describing some rheometers designed for simplicity of construction and operation, and then go on to more complex instruments designed for greater precision and versatility. A good example of the former approach is the concentric cylinder creep-meter of Dexter [140], shown in Figure 5-2. A fixed torque is applied to the inner cylinder (5) by a dead weight (3) acting through pulleys (1, 2), and the rotational displacement is recorded by means of a

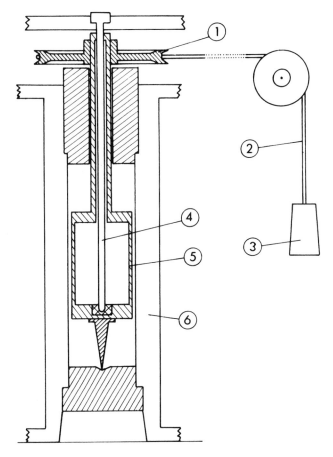

Figure 5-2. Concentric cylinder rheometer of Dexter [140]. (Reproduced from J. L. Leblanc [16] with permission of Editions CEBEDOC, Liege.)

mechanical linkage that drives a pen on a chart recorder. There are no electronic components, and electricity is required only to power the chart paper drive. Pulley friction and the inertia of the moving components, especially the dead weight, are obvious sources of error.

An even simpler device is the "rotational shear plastometer" designed by McCord and Maxwell [141] to measure elastic recoil. The inner cylinder is displaced manually through a fixed angle and then released. The recoil is indicated by a pointer attached to the top of the shaft of the inner cylinder. This is a relatively wide gap instrument and thus subject to a number of the errors mentioned in the previous section, but the authors intended the instrument only for the "relative comparison of various materials." A more sophisticated version of this apparatus, the "melt elasticity tester," has been described by Maxwell and Nguyen [142].

Aloisio et al. [143] measured the relaxation function, $G(t)$, of molten polymers using a device having some similarity to the one described above. To provide a large shearing surface, a double concentric cylinder arrangement was used in which an inverted cup rotates in an annular shaped melt container. Thus, shearing stress acts on both sides of the cup walls. The gaps on the two sides have different dimensions, and the small gap approximation was used to derive a relationship between $G(t)$ and the measured torque and angular displacement.

Until fairly recently, reliable electronic torque transducers able to operate in conjunction with relatively rigid rotating shafts were not available, and torsion wires or rods were used for torque measurement. For example, Kepes [138] designed a melt rheometer for viscosity measurement in which a torsion wire was used. Guard rings were used to minimize end effects.

When time-dependent properties are to be measured, the elasticity of the wire itself and the moment of inertia of the cylinder complicate the interpretation of the data. Den Otter [144] measured the storage and loss moduli of melts by use of a "forced torsional pendulum" in which the upper end of the torsion wire was oscillated while the amplitude and phase shift of the inner cylinder suspended from it were observed. Horio et al. [145, 146] and Vinogradov et al. [147, 148] used rheometers in which the outer cylinder was oscillated and the inner cylinder was supported on a torsion wire.

Rokudai and Fujiki [149] designed a rheometer to study stress relaxation after cessation of the steady shear generated by the rotation of the outer cylinder. The deflection of the torsion wire is monitored by means of a sensitive angular displacement transducer, and the signal so generated is fed to a servo motor, which rotates the torsion wire mounting so as to maintain the inner cylinder stationary.

The recent trend has been toward the use of torque transducers that operate in conjunction with shafts and moment arms that are rigid compared to the material under study so that their deflection can be neglected in the analysis of data. Dealy *et al.* [150] used a torque transducer which operates on the basis of magnetic stress anisotropy in a steel tube. This device produces a large signal and is thus relatively immune to noise. It has a high overload capacity and a linear characteristic, and is robust and reliable. The rheometer itself incorporated cylinders with slightly tapered walls so that the gap spacing could be altered simply by raising the inner cylinder. The outer cylinder could either be rotated at constant speed or oscillated by means of an eccentric operating through a rack and pinion, while the torque on the inner cylinder was measured. Thermostatting oil was circulated both inside the inner cylinder and outside the outer cylinder. Tee and Dealy [29] used this rheometer to study the response of melts to large amplitude oscillatory shear.

McCarthy [43] subjected melts to large amplitude oscillatory shear in a rheometer in which the gap is filled by means of a unique "ring extruder" system that avoids the use of premolded samples.

In using rotational rheometers to study the effect of pressure on viscosity, the major problem is the measurement of torque. In the case of low viscosity liquids, magnetic couplings are often used to transmit torque to a rotating cylinder in a sealed chamber, but this technique has not been used for melts. Semjonov [151, 152], Cogswell [153], and Christmann and Knappe [154] built high pressure rheometers in which the outer cylinder was rotated and the torque was inferred from the torsional deflection of a hollow tube, on the end of which was mounted the inner cylinder. Cogswell [153] used a ram fitted into the outer cylinder to pressurize the melt.

In a hollow shaft sensor, the angular deflection of a torque tube located inside the high pressure test chamber is indicated by means of a rod fastened to the rheometer cylinder and passing out of the test chamber through the torque tube. H. K. Bruss* manufactures a series of industrial process viscometers which make use of this technique for

---

*Address given in Appendix B. Some aspects of the Bruss viscometers are covered by U.S. Patent 2,518,378.

viscosity measurement up to 450°C and 300 atm. The output is an electronic signal proportional to torque, and in some models Bruss has adapted the hollow shaft sensor for use with the rotating member of the viscometer.

Another approach to the problem of high pressure rheometry is that of Kuss [155], who rotated the inner cylinder by means of a shaft passing through high pressure seals, while the torque on the outer cylinder was measured by means of a transducer mounted inside the high pressure chamber.

Before closing this section, mention should be made of an unusual rheometer designed to be an integral part of a single screw extruder. Revesz [156] mounted a torsion sensitive tip on the end of the screw. He suggests a method for estimating the viscosity of the melt as it enters the die, but the principal purpose of the device is to provide a basis for the control of the extrusion rate.

5.2 CONE AND PLATE FLOW

The flow that is most used to measure the viscometric and linear viscoelastic material functions is the flow between a cone and a plate, one of which is rotating. A sketch of this geometry is shown in Figure 5-3. The principal advantages of this geometry are that loading and cleaning are relatively easy and the shear rate is uniform, so that differentiation of data is not necessary to compute values of material functions.

On the other hand, flow irregularities at the air-liquid interface are

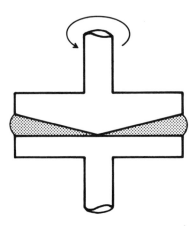

Figure 5-3. Cone and plate geometry.

especially severe for molten polymers. Thus, the use of cone and plate rheometers to study these materials is limited to a maximum shear rate in the range of 0.1 to 10 s^{-1}.

5.2.1 Basic Equations

In order to derive a simple expression for the shear rate in cone and plate geometry, it is necessary to make a number of assumptions and approximations, and these are valid only for certain combinations of cone angle, rotational speed and fluid properties. Under these circumstances, the flow is "approximately controllable." This means that the shear rate can be specified within an acceptable margin of error. When these assumptions and approximations are not valid, the flow may be rather complex and the cone plate geometry ceases to be of use for measuring rheological properties.

To derive an expression relating the shear rate in the space between the cone and the plate to the rotational speed and cone angle, the following assumptions are made.

1. At sufficiently low rotational speeds, the "inertia" or acceleration terms of the equation of motion can be neglected, and the flow predicted by the remaining terms will differ only marginally from the exact solution of the entire equation.
2. The cone angle is very small so that certain approximate trigonometric relationships can be used to simplify equations.
3. The free surface of liquid at the edge of the gap between the cone and the plate is spherical in shape with a radius of curvature equal to the cone radius, and the flow is uniform right up to this surface. In other words, edge effects are neglected.
4. The surface tension operating at the free surface has no effect on the fluid motion.

When these assumptions are valid, the shear rate is approximately uniform and is given by the following expression*:

$$\dot{\gamma} = \frac{\sin\theta}{r} \frac{\partial}{\partial\theta}\left(\frac{V_\phi}{\sin\theta}\right) = \frac{\Omega}{\theta_0} \qquad (5\text{-}15)$$

---

*For a derivation of this equation, see Walters [14].

Since the shear rate is nearly uniform, the shear stress will also be uniform, and the torque required to rotate the cone (or to hold the plate stationary) is given by

$$M = \int_0^R \tau_{\theta\phi} 2\pi r^2 dr = \tfrac{2}{3} \pi R^3 \tau_{\theta\phi}. \qquad (5\text{-}16)$$

It is thus possible to compute the value of the viscosity at one shear rate from the results of a single experiment.

$$\eta = \frac{\tau_{\theta\phi}}{\dot{\gamma}} = \frac{3M\theta_0}{2\pi R^3 \Omega} \qquad (5\text{-}17)$$

If this geometry is to be used to determine the linear viscoelastic functions, G' and G'' by oscillating one member with an angular amplitude, ϕ, and measuring the torque amplitude, M, and the loss angle, δ, the following equations can be employed.

$$G' = \frac{3\theta_0 M_0 \cos\delta}{2\pi R^3 \phi_0} \qquad (5\text{-}18)$$

$$G'' = \frac{3\theta_0 M_0 \sin\delta}{2\pi R^3 \phi_0} \qquad (5\text{-}19)$$

Turning now to the normal stress differences, these can be related to the wall pressure at the surface of the plate. We will use the symbol ΔP here to indicate the wall pressure less the ambient pressure.

$$\Delta P \equiv P_w - P_a \qquad (5\text{-}20)$$

It can be shown that the gradient of this quantity with r is related to the normal stress differences as follows.

$$r \frac{d(\Delta P)}{dr} = - \left[N_1(\dot{\gamma}) + 2N_2(\dot{\gamma}) \right] \qquad (5\text{-}21)$$

At the free surface, the "rim pressure," $P_w(R)$, is related to the second normal stress difference as shown in Equation 5-22.

$$\Delta P(R) = - N_2 \qquad (5\text{-}22)$$

Equation 5-21 can be integrated, and the result combined with Equation 5-22 to yield the following expression for the total normal force on the plate.*

$$F = \int_0^R 2\pi r(\Delta P)\,dr = \frac{\pi R^2}{2}\, N_1 \qquad (5\text{-}23)$$

Equation 5-21 implies that if P_w is plotted as a function of $\ln(r)$, a straight line will result, the slope of which is related to the quantity ($N_1 + 2N_2$). Then Equation 5-22 can be used to calculate N_1 from rim pressure measurements so that, in principle, both N_1 and N_2 can be determined from the wall pressure distribution. However, the measurement of $P_w(r)$ requires the installation of a series of very small flush-mounted pressure transducers in the plate. The measurement of the total normal force, on the other hand, does not require such transducers, although it yields no information regarding the second normal stress difference.

Brindley and Broadbent [157] and Miller and Christiansen [158] used planar transistors as pressure transducers to determine the wall pressure distribution on the plate for polymer solutions. The former workers found the rim pressure method of calculating N_2 unreliable, but the second team found good agreement betwen N_2 values obtained by use of Equation 5-22 and those obtained by use of Equations 5-21 and 5-23. However, both sets of authors found that the second normal stress difference is much smaller than the first. Only the total normal force method appears to have been used to date for melts.

The derivation of Equations 5-21 through 5-23 is based on the assumption that the inertia (acceleration) terms in the equations of motion are negligible. It is known that if these terms are included, the Newtonian fluid equations predict secondary flow, and the effect of this on the use of cone plate rheometers will be discussed in the next section. However, even in the absence of secondary flow, the inertia terms can make a contribution to the wall pressure. This problem has been examined by Walters (p. 63 of [14]), who presents equations for the wall pressure and normal force, including the effect of inertia, but still assuming negligible secondary flow. If N_1 is small, failure to take this

---

*There are two minor errors in the derivation of this equation given by Walters [14].

effect into account can lead to the calculation of a negative value of N_1, but when the first normal stress difference is large, as it usually is for polymer melts, the inertial correction is insignificant.

5.2.2 Sources of Error in Cone and Plate Flow

Viscous heating and temperature fluctuations. The gap spacing, and thus the temperature distribution across the gap, depends on the radial position. Bird and Turian [159] found an approximate solution for the case of a Newtonian fluid with constant viscosity when the surfaces of the cone and plate are maintained at a fixed temperature, T_w, and there is no heat transfer across the free surface. The maximum deviation of the steady state fluid temperature from T_w is as follows.

$$T_{max} - T_0 = \frac{3 M \Omega \theta_0}{16 \pi k R} = \frac{\eta \Omega^2 R^2}{8 k} \qquad (5\text{-}24)$$

However, in the case of melts the speed is limited to rather low values by the onset of edge irregularities, and temperature nonuniformity is usually not a serious problem.

Variations of temperature with time resulting from the operation of the temperature control system, however, can introduce significant errors into normal force measurements. These variations will result in fluctuations in system geometry and sample density, which will have associated with them a small pulsatile flow of polymer in and out of the gap and a consequent contribution to the normal force. This error will be most severe in the case of fluids with large viscosity and small normal stress differences.

Secondary flow. Cone and plate flow is only approximately controllable, which means that the shear rate given by Equation 5-15 is an approximation valid only when acceleration effects are small. Secondary flow, in the form of toroidal vortices, is present in Newtonian fluids even at low rotational speeds [160], but it has little effect on torque and normal force under these conditions. As the speed is increased, the strength of the vortex flow becomes greater and will eventually result in substantial errors in calculated property values.

Cheng [161] measured the effect of secondary flow on the accuracy of viscosity values for a Newtonian fluid, and Turian [162] found good agreement between these data and his theoretical treatment of the flow, which predicts, for example, that a 10% deviation in the torque, and thus in the calculated viscosity, will result when the Reynolds number is about 13.

Details of the secondary flow motion have been studied by Savins and Metzner [163] and by Kulicke and Porter [164] for polymer solutions, and by Ballman *et al.* [165] for melts. The former authors argue that secondary flow has a much stronger effect on the normal force than on the torque, and they recommend that in order to keep the effect of secondary flow on F to a negligible level, the Reynolds number must be less than 0.5. The error in calculated values for N_1 caused by secondary flow depends on the rheological properties of the fluid so that no simple generalization based only on Reynolds number is possible.

The use of small cone angles will minimize the error due to secondary flow, but for melts, edge effects usually become troublesome at a speed well below that at which this error becomes significant.

Shear rate nonuniformity. Equations 5-15 and 5-17 are approximations valid for small cone angles, and the use of finite cone angles will always lead to some variation in shear rate across the gap. Adams and Lodge [32] have calculated the effect of this variation for Newtonian liquids, and they found that for a cone angle of 10°, the variation in shear rate is 3% and the resulting error in calculated viscosity is 2%. Since this is the largest cone angle ever used in rheometry, shear rate nonuniformity is not thought to be a significant source of error. Paddon and Walters [166] have shown that shear rate nonuniformity is greater for non-Newtonian fluids, but that for cone angles equal to or less than 4°, the resulting error is quite small.

Edge effects. One of the assumptions used to derive the rheometrical equations for cone plate flow is that the free surface is spherical with a radius of curvature equal to the cone radius. Among other things, this presumes that the volume of liquid loaded into the gap is exactly the appropriate amount. The torque and the gradient in wall pressure are not strongly affected by the detailed shape of the free surface, but this may have an important effect on the rim pressure and thus on the normal force, especially when surface tension forces are significant compared to elastic forces. In polymer melts, the elastic forces normally predominate over the surface tension forces, so that small deviations in the shape of the free surface are not expected to be a serious problem.

However, at sufficiently large rotational speeds, the fluid deformation can cause a pronounced change in the shape of the free surface, as shown in Figure 5-4. The fluid tends to flow outward near the walls of the cone and the plate. The torque decreases markedly and then fluctuates. As speed is further increased, air bubbles may be entrapped when the speed is reduced again, making it impossible to reproduce low shear rate data when the shear rate is reduced. Hutton [167] was among the first to report on this phenomenon, and has suggested [168] that it is the result of "fracture" along a conical surface and that its severity increases as the product $(N_1 R \theta_0)$ increases. There have been several subsequent reports of "edge failure" in cone and plate flow [169, 170, 171], and it is now recognized that this phenomenon governs the maximum shear rate at which cone and plate geometry can be used in the study of molten polymers. C. D. Han has suggested to the author that this upper limit can be most simply expressed in terms of a maximum shear stress and that this maximum value is about 5×10^4 Pa. This criterion, however, makes no reference to normal stress difference and cone angle, which are the quantities emphasized by Hutton [168]. Clearly, much remains to be learned about the detailed nature of "edge fracture."

Nonideal geometry. Deviations in the flow geometry from that for a perfectly shaped, aligned and positioned cone and plate will lead to errors in measured properties. First, the apex of the cone is assumed to be just touching the surface of the plate, and there are several mechanisms that can cause deviations from this ideal situation; for example, lack of care in the positioning of the movable member after a sample has been inserted. Even if the gap is set precisely at room temperature, thermal expansion will usually require repositioning at operating temperature.

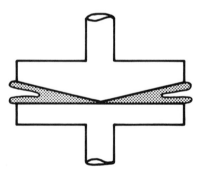

Figure 5-4. Distortion of free surface at high shear rate.

An additional mechanism that can alter the gap spacing once the rheometer has been put into operation is the deflection of the rheometer frame (or the normal force sensing aparatus) in response to a large, positive first normal stress difference. If the instrument compliance is sufficiently large, this can lead to a significant increase in the gap spacing, so a high degree of stiffness must be built into the rheometer. Brindley and Keene [172] carried out a theoretical analysis of the effect of lack of rigidity on the oscillatory testing of stiff materials.

When the normal force is small—for example, at low shear rates—a periodic oscillation of the force signal is often observed, and Adams and Lodge [32] have attributed this to slight imperfections in the bearings, which give rise to periodic fluctuations in the gap spacing. If the amplitude of these fluctuations is Δh_1, they estimate the amplitude, P_0, of the resulting wall pressure to be proportional to the following combination of parameters.

$$P_0(r) - P_w(R) \propto \left[\frac{-6\eta\Omega\Delta h}{\theta_0^3} \left(\frac{1}{r} - \frac{1}{R} \right) \right] \qquad (5\text{-}26)$$

Clearly, this problem is most severe when the cone angle is very small.

Adams and Lodge [32] also examined the effect of "tilt"; i.e., the deviation from 90° of the angle between the axis of symmetry of the cone and the plane of the plate, and found that the resulting error in $\Delta P(r)$ cannot be completely eliminated by averaging results obtained by rotation in both directions.

Finally, one needs to verify the geometry of the cone itself. Cheng and Davis [173] carefully examined cones supplied by two rheometer manufacturers and found surprising discrepancies between the actual and specified dimensions and shapes.

Selection of cone angle and radius to minimize errors. An important aspect of the errors resulting from geometrical nonidealities is that they increase as cone angle decreases, whereas all the other errors mentioned above diminish as the cone angle decreases. Meissner [174] found it necessary to use cone angles in the range of 8° to 10° to obtain stress growth results that were independent of cone angle, and Crawley and Graessley [175] observed this independence in the range from 4° to 8° but not at smaller angles. These observations probably result from systematic errors associated with instrument compliance. However, to extend the range of measurement to higher shear rates, it is necessary to decrease the cone angle and radius to keep viscous heating and edge failure under control.

As a general rule, extreme caution should be exercised in the use of cone and plate geometry at shear rates above 10 s^{-1}, and even the range between 1 and 10 s^{-1} results can be subject to large errors for some materials.

5.2.3 Some Cone Plate Rheometers for Molten Polymers

The cone is nearly always placed above the plate, but one may choose to rotate either member. In the rheometers of Trapeznikov and Pylaeva [176] and Raha et al. [177], it is the cone that rotates. In the latter instrument, a set of ten synchronous motors is used to rotate the cone.

Benbow [97] designed a constant stress viscoelastometer that consisted of a cone plate rheometer in which the torque was applied to the cone by means of a weight on the end of a filament passing over a pulley and wrapping onto a drum attached to the shaft of the cone. This arrangement permits the determination of viscosity and recoverable shear.* Benbow and Lamb [86] used a miniaturized version of this instrument for use with very small samples.

In the "microconsistometer" of Kepes [138] and the "elastoviscometer" of Vinogradov and Belkin [178], the rotating plate is part of a cup which is filled with liquid up to the level of the rim of the cone. This, of course, is a deviation from the edge condition assumed in deriving the rheometrical equations, and Walters ([14], p. 67) has discussed the implications of using the "sea-of-liquid" system. In the Kepes instrument, a guard ring is used to reduce the edge effect, and a series of torsion wires of varying stiffness are used to measure the torque. This instrument is designed for use with very small samples. Komuro et al. [179] also used the guard ring concept in their "Rion" cone plate rheometer.**

The "elastoviscometer" of Vinogradov and Belkin [178] is shown in Figure 5-5. It is designed for operation under vacuum, and the torque transducer (6) mounted on the rigid cone shaft (4) is located inside the

---

*A commercial version, the Davenport cone and plate viscometer, is described in Section 9.8.1.
**A commercial rheometer formerly manufactured in Japan.

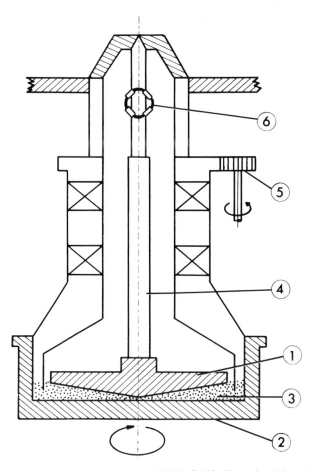

Figure 5-5. Elastoviscometer of Vinogradov and Belkin [178]. (Reproduced from J. L. Leblanc [16] with permission of Editions CEBEDOC, Liege.)

viscometer chamber. The lower plate (2) is an integral part of this chamber and is rotated by a gear mechanism (5).

Raha [180] designed a melt rheometer in which the plate undergoes forced rotational oscillation and the cone is mounted on a torsion rod. Equations attributed to Weissenberg are used to calculate $\eta'(\omega)$ and $G'(\omega)$. Batchelor [181] used a cone and plate rheometer to study elastic recoil, with a capacitor used to measure the rotational displacement. Menefee [182] designed a cone plate rheometer for use in the study of creep, creep recovery and relaxation after cessation of steady shear.

Most of the measurements based on cone and plate geometry that were carried out in the 20-year period starting around 1955 made use of a rheometer called the Weissenberg Rheogoniometer.* This is the commercial version of the first rheometer designed to measure both shear and normal stresses. Weissenberg's original design [183] was later improved by Roberts [184, 185], and the improved version is sometimes referred to as the "Weissenberg-Roberts Rheogoniometer."

Some of the main features of the Weissenberg Rheogoniometer are shown in Figure 5-6. The cone (4) is mounted on a torsion bar (1), and its angular displacement is monitored electronically (2). The plate (6) is driven by a mechanical mechanism (7), and the normal force on it is also monitored electronically (8). The temperature of the sample (5) is controlled by means of a radiation oven (3).

The normal force detector on this instrument is of such a design that significant axial deflection can occur when highly elastic materials are studied. For this reason, a servo system is provided to minimize the resulting deflection of the cone, but the response time of the servo system makes it unsatisfactory for transient experiments such as stress growth. In addition, users interested in melts having a high first normal stress difference have found the rheometer frame to be too compliant. For these reasons, a number of users [174, 175, 186, 187, 188, 189, 190] have modified their instruments to improve the suitability for the study of highly elastic liquids. Higman [191] developed a torque and normal force measurement system based on the use of a single piezo-electric crystal.** Meissner [192] further modified his Weissenberg Rheogoniometer by adding a servo-controlled drive motor for the plate. This makes possible the use of a wide variety of test modes including creep and elastic recovery. Menezes and Graessley [193] added a high speed data acquisition system and mounted a shaft encoder on the platen. The latter provides a digitally encoded position signal which can be fed directly to a microprocessor.

Hayashida and Okuda [194] used a cone plate rheometer in which polymer was continuously injected around the edge by means of an

---

*Further information on the commercial version of the Weissenberg Rheogoniometer is given in Section 10.2.
**A commercial version of this system, made by Deimos, is described in Section 10.2.

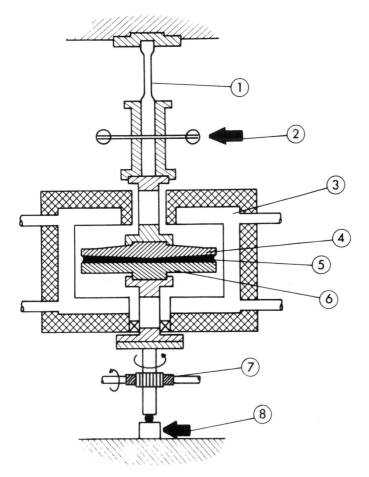

Figure 5-6. The Weissenberg Rheogoniometer. (Reproduced from J. L. Leblanc [16] with permission of Editions CEBEDOC, Liege.)

extruder and allowed to flow out through a hole in the center of the plate under the influence of the normal stress gradient. The extruder feed rate was adjusted to maintain ambient pressure at the rim. The continuous flow of polymer through the rheometer limited the extent of thermal degradation, and first and second normal stress differences were reported at shear rates as high as 220 s^{-1}. It was assumed that the usual rheometric equations were still valid in spite of the radial component of velocity.

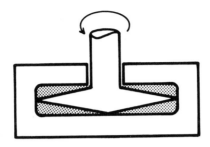

Figure 5-7. Biconical geometry.

5.2.4 The Biconical Rheometer

The biconical geometry is illustrated in Figure 5-7. It has the advantage of eliminating the free surface and thus the variation of its shape with rotational speed. It also minimizes the exposure of the sample to the environment. However, a new type of edge effect is introduced by the presence of the cavity side walls. Also, secondary flows are no less troublesome than in cone and plate flow.

This geometry has been used by Piper and Scott [195] to study rubbers and by Pohl and Gogos [196] and by Madonia and Gogos [197] to study molten thermoplastics. Best and Rosen [198] built a simple, dead-weight loaded biconical creepmeter for melts. Broyer and Macosko [199] compared results obtained with several rotational flows including biconical flow. They presented equations for correcting the measurements to take account of the cavity walls and the shaft hole. Gavin and Whorlow [171] also compared the results of a biconical rheometer with those of several other melt rheometers and found the data satisfactory at shear rates up to several reciprocal seconds. They verified a formula originally given by Piper and Scott [195] for calculating the edge effect.

5.2.5 Extended Cone and Plate Flow

Jackson and Kaye [200] pointed out that if the cone apex is separated from the plate by a distance, h, the flow is still approximately viscometric, and making assumptions similar to those made in the case of cone and plate flow, they derived the following rheometrical equations.

$$\dot{\gamma} = \frac{\Omega R}{h + R\theta_0} \tag{5-27}$$

$$F = \pi \int_0^R \left[N_1 - \frac{N_2 h}{h + R\theta_0} \right] r dr \tag{5-28}$$

$$N_2(\dot{\gamma}_0) = -\frac{\theta_0}{\pi R} \left(\frac{\partial F}{\partial h} \right)_{h=0} \Omega - \dot{\gamma}_0 \left(\frac{dN_1}{d\dot{\gamma}} \right)_{\dot{\gamma}_0} \tag{5-29}$$

The interesting feature of this flow is that the relatively easily measured normal force depends on the second normal stress difference. Thus, by making measurements at various values of h, including $h = 0$, one might hope to obtain a more precise value of the second normal stress difference. Petersen et al. [201] have used this geometry to study molten polymers. However, they found that N_2 was so small that it was difficult to distinguish it from the effect of secondary flow.

5.3 FLOW BETWEEN ROTATING PARALLEL PLATES

In this flow, sometimes called "torsional flow," the fluid is placed between two parallel circular disks of radius R, one of which rotates with an angular velocity Ω. Figure 5-8 shows this geometry.

Making simplifying assumptions similar to those made in the case of cone and plate flow (see [147]), one can derive the following equations.

$$\dot{\gamma} = r\Omega/h \tag{5-30}$$

$$\frac{2F}{\pi R^2} \left[1 + \frac{1}{2} \frac{d\ell n F}{d\ell n \dot{\gamma}_R} \right] = N_1(\dot{\gamma}_R) - N_2(\dot{\gamma}_R) \tag{5-31}$$

$$\eta(\dot{\gamma}_R) = \frac{3M}{2\pi R^3 \dot{\gamma}_R} \left(1 + \frac{1}{3} \frac{d\ell n M}{d\ell n \dot{\gamma}_R} \right) \tag{5-32}$$

As can be seen from Equation 5-30, the shear rate in the gap is not even approximately uniform. This makes it impossible to calculate values of material functions on the basis of a single experiment, and differentiation of data is required as indicated by Equations 5-31 and 5-32. Since the normal force depends on both the first and second normal stress differences, it is not possible to determine their individual values from

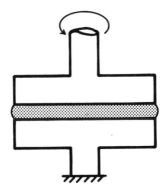

Figure 5-8. Parallel plate (torsional flow) geometry.

normal force measurements. However, if N_1 has been previously determined using cone and plate flow, then N_2 can be determined using the parallel plate flow.

The sources of error for torsional flow are the same as for cone and plate flow, although secondary flows are not promoted by the presence of normal stress differences in this case.

Sakamoto *et al.* [202] used a rheometer designed by Kotaka *et al.* [203] to determine the wall pressure distribution for polyethylene. Sakamoto and Porter [204] used parallel plate fixtures installed in a Weissenberg Rheogoniometer, and Broyer and Macosko [199] compared data from this geometry with those obtained using cone and plate and biconical geometries.

Plazek [205] designed a parallel disk rheometer with virtually no friction for use in studying creep and creep recovery. Torque is applied to the upper surface of the sample by a rotor suspended in a magnetic field and driven by a drag cup motor. The sample is in a sealed chamber so that its environment can be readily controlled. The parallel disk geometry is especially convenient for the study of melts, because samples can be cut from sheets.

Binding and Walters [206] have used a "torsional balance rheometer" based on the parallel disk flow. However, instead of measuring the normal force resulting from a fixed plate spacing, they apply a known load to the upper plate and measure the steady-state plate spacing.

If torsional flow is to be used to determine linear viscoelastic properties by oscillating one plate with an angular amplitude, ϕ_0, and measuring the torque amplitude, M_0, and its phase lag, δ, the following relationships can be used.

$$G' = \frac{2 M_0 h}{\pi R^4 \phi_0} \cos \delta \qquad (5\text{-}33)$$

$$G'' = \frac{2 M_0 h}{\pi R^4 \phi_0} \sin \delta \qquad (5\text{-}34)$$

5.4 MOONEY VISCOMETER FLOW

This flow, first used by Mooney [207], occurs between a disk-shaped rotor and a geometrically similar cavity. It has often been used to characterize elastomers, and ASTM standard test method D 1646-74 is based on it.* The ASTM standard specifies a cavity diameter of 50.93 millimeters. The test procedure calls for the torque to be measured at a speed of 2 RPM and reported as "Mooney viscosity" (ML), defined to be 100 when the torque is 8.3 newton-meters.

White and Tokita [208] used this geometry as the basis for a variable speed rheometer, and they derived equations for calculation of the viscosity and shear rate by neglecting inertia, fluid elasticity and edge effects. Nakajima and Collins [209] used these equations to interpret transient data obtained in the study of some elastomers and compared the results with properties measured using other techniques.

Nakajima and Harrell [210] considered the role of edge effects and showed that the viscosity calculated by use of the simplified formulas is somewhat larger than the true value. They derived new rheometrical equations for the case of a power law fluid.

Mooney rheometer flow is subject to many sources of error, and its usefulness is limited to quality control and material comparisons. Its use as a melt rheometer is limited to the measurement of viscosity at low shear rates.

5.5 ECCENTRIC ROTATING DISK ("ERD") FLOW

This flow is related to oscillatory shear and is used to determine the linear viscoelastic properties, $\eta'(\omega)$ and $G'(\omega)$. It is a curious flow, how-

---

*Commercial versions are described in Section 10.8.2. The corresponding European tests are DIN 53523, BS 1673 and ISP R 289.

ever, in that it is a "flow with a constant stretch history," as defined in Chapter 2. First let's see how this particular deformation is generated in the laboratory and then try to visualize it from the point of view of a fluid element.

The fluid is placed between two parallel disks, both of which can rotate about their axes. These axes are not colinear, but are offset by a distance d. This arrangement is shown in Figure 5-9. The upper (or lower) plate is driven at a rotational speed, Ω, and the shaft of the other plate turns in a bearing with sufficiently small friction that there is virtually no resistance to its being dragged along by the fluid at a speed essentially equal to Ω. The lateral forces exerted by the fluid on the lower plate are measured, and we will see that this permits the calculation of η' and G' at a frequency equal to Ω.

Now we turn to the question of how dynamic properties can be determined in a steady motion. Without delving into the mathematical analysis of this situation, it is instructive to visualize the deformation pattern as "seen" by an element of fluid. The following procedure is recommended. Hold your hands out about 2 centimeters apart, with the palms facing each other and the fingers parallel. Imagine that the fluid fills the space between your hands. Now slide your right hand* about 2 centimeters toward your body, keeping the spacing constant and the fingers parallel. (This provides the offset.) Now move the right hand in a circle,

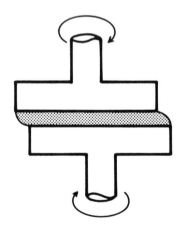

Figure 5-9. Eccentric rotating disks. (Maxwell orthogonal rheometer.)

---

*If you are left-handed, you may prefer to offset and move the left hand.

keeping the spacing constant *and the fingers parallel*. (If you find it impossible to do this without amputating one or both hands, you are doing something wrong.) What we have done here is to look at the deformation undergone by a fluid element without the superposed and misleading bulk rotation of the fluid element. This bulk rotation is rheologically irrelevant, and we see that the rotation of the plates is simply a mechanism for providing the desired deformation.

If you look at the edge of your hands as you carry out this demonstration, you will see something which looks very much like the oscillating plate configuration used to describe oscillatory shear in Section 2.4.4. The only difference is that, in this case, the plate is not simply oscillating along a line, but moving around in a circle. It is *not* rotating; the rotation is introduced because we choose to generate the deformation by the use of eccentric rotating disks.

This deformation pattern had been used for the study of rubbers, but its use for the study of molten plastics was pioneered by Maxwell [211, 212]. He called this device an "orthogonal rheometer," and the deformation pattern is often referred to as "Maxwell orthogonal rheometer flow."*

5.5.1 Basic Equations for Eccentric Rotating Disks

Walters ([14], p. 168) has given a derivation of the rheometrical equations for this flow, and only the results will be given here. The derivation is based on the assumption that the strain amplitude, $\gamma_0 = d/h$, is small and that the inertia (acceleration) terms in the equations of motion can be neglected. (Possible errors resulting from deviations from this ideal situation are discussed in the next section.) The x and y directions are defined in Figure 5-10, and if lateral forces, F_x and F_y, exerted by the fluid on the lower plate, can be measured, the dynamic viscosity, η', and the storage modulus, G', can be calculated as follows.

$$\eta' = \frac{F_x h}{\pi \Omega d R^2} \tag{5-33}$$

$$G' = \frac{F_y h}{\pi d R^2} \tag{5-34}$$

The frequency, ω, is equal to the angular speed, Ω.

---

*A commercial instrument, based on the Maxwell design, is described in Chapter 10.

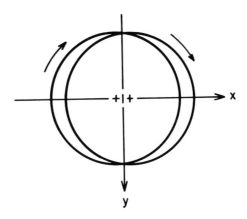

Figure 5-10. Top view of eccentric rotating disks.

There is also a normal thrust, F_z, and Macosko and Davis [213] have examined the possibility of using this quantity to calculate the ratio N_2/N_1.

5.5.2 Sources of Error for ERD Flow

Effect of strain amplitude on linearity. Equations 5-33 and 5-34 imply that, at a given speed, Ω, the forces are proportional to the strain amplitude, γ_0, defined as

$$\gamma_0 \equiv d/h.$$

This is because η' and G' depend only on ω (which is equal to Ω). These equations are based on the assumption that γ_0 is very small, and there will always be some value of γ_0 beyond which the forces will no longer be proportional to it so that Equations 5-33 and 5-34 cannot be used. This "limit of linear viscoelastic response" has been studied experimentally by Gross and Maxwell [214] and by Macosko and Davis [213]. It appears that the "viscous" force, F_x, remains proportional to angular frequency at considerably larger values of γ_0 than the "elastic" force, F_y. Thus, Equation 5-33 has been found to be valid for a number of melts at strains as high as 1.0, while Equation 5-34 fails at strains above 0.5.

Jongschaap *et al.* [215] carried out a theoretical analysis of the nonlinearity of ERD flow and concluded that, at sufficiently low frequencies, the limiting strain amplitude is inversely proportional to the square of the frequency, while at high frequencies, the limiting strain becomes independent of frequency. The experimental results mentioned above indicate that the limiting strain is independent of strain, and this implies that frequencies of practical interest are "high" as regards this theory.

Rather than relying on a rule of thumb, it is, of course, always best to demonstrate linearity for the particular material under study by verifying that calculated values of η' and G' are independent of (d/h) at a fixed value of Ω.

Inertia. It will be recalled that inertia was neglected in deriving Equations 5-33 and 5-34. Abbott and Walters [216] showed how these equations are altered when inertia is taken into account.* Macosko and Davis [213], however, concluded from their experimental study that inertial effects are not important for high viscosity fluids such as polymer melts. Of course, at a sufficiently large angular speed liquid will be thrown out of the gap.

Speed lag. The rheometrical equations for ERD flow depend on the assumption that both disks rotate at the same speed. However, it has been shown by Payvar and Tanner [217] that a certain amount of slip is bound to occur when the offset, d, is not zero, even if the bottom (freely rotating) disk is mounted in a perfectly frictionless bearing. For a Newtonian fluid, they found that the angular velocity, Ω_1, of the freely turning disk is related to that for the driven disk, Ω_2, as follows.

$$\Omega_1 = \Omega_2[1 - 2(d/R)^2] \qquad (5\text{-}35)$$

Davis and Macosko [218] made an experimental study of lag for high viscosity fluids. They found that the use of an air bearing introduced insignificant lag above the inevitable hydrodynamic lag described above. For fluids having viscosities above 100 Pa·s, the lag was less than 0.5% of Ω.

Edge effects. The flow field in the neighborhood of the air-liquid interface at the edge of the disks is unlikely to be homogeneous, and this could contribute to inaccurate values of the material functions. However, based on the experimental studies of Payvar and Tanner [217] and Macosko and Davis [213], edge effects do not seem to be a significant source of error.

---

*This work is also reported by Walters ([14], p. 18).

Viscous dissipation and temperature nonuniformity. Ahrens and Goldstein [219] have calculated the temperature distribution resulting from viscous dissipation in ERD flow. When inertia is negligible, the maximum temperature rise is as follows.

$$T_{max} - T_w = \eta' R^2 \Omega^2 / 2k \qquad (5\text{-}36)$$

This equation is valid when

$$\frac{\Omega \rho h^2}{2\eta'} \ll 1.$$

When this condition is not satisfied, the temperature rise is dependent not only on this parameter, but also on the elasticity of the liquid. However, polymer melts are sufficiently viscous that inertia and elasticity may be neglected in estimating the maximum temperature rise.

Instrument compliance. In addition to the lateral forces, F_x, and F_y, which are measured to determine η' and G', there is a normal component of force, F_z, and this will result in an axial force on the shafts of the two disks. If the rheometer body is not sufficiently stiff, this can result in a significant change in the gap spacing, h, and thus an error in the calculated properties.

Macosko and Davis [213] studied the effect of instrument compliance as well as that of tilt.

5.6 OTHER "ECCENTRIC" ROTATIONAL FLOWS

A number of flows are similar to ERD flow in that they are time-steady flows which, nonetheless, can be used to determine the linear viscoelastic properties. One example is the flow between concentric hemispheres rotating about axes that are at a small angle one to the other. Képès [220] developed a rheometer based on this flow. A commercial version of this instrument, the Képès Balance Rheometer, is made by Contraves. The maximum operating temperature is only 100°C, so it is of limited utility for the study of melts, although the author has seen such an instrument modified for use with melts at temperatures up to 300°C. Resin pellets can be loaded directly into the lower hemisphere and melted in place.

Other such geometries are eccentric rotating cylinders, displaced spheres, tilted cylinders and tilted cone and plate. Walters [14] and Macosko and Davis [213] have discussed some of these flows. For various reasons, these do not appear to be as useful for the study of melts as eccentric rotating disk flow.

Chapter 6
Uniform Extensional Flows

This chapter describes the several uniform stretching deformations that have been used to determine well-defined material functions for molten polymers. Other stretching flows that have been used to study melts are converging flow, melt spinning and film blowing. However, these deformations are not uniform in space, and there is no unambiguous way to interpret the data in terms of the material functions described in Section 2.3.2. Their use in the characterization and comparison of plastic materials is described in Chapter 7.

Certain flows, such as the tubeless siphon (Fano flow) and the triple jet, which have been used in the study of polymer solutions, are not suitable for use with high viscosity melts and are not considered in this book.

Several difficulties arise in the experimental study of uniform extensional flows that have no counterpart in the study of uniform shear flows. First, the sample must be supported in such a way as to eliminate gravity as a driving force for deformation, without hindering the elongational process. This problem does not arise in the case of extremely viscous materials with viscosities higher than about 10^7 Pa·s, but for thermoplastics at typical processing temperatures, it cannot be ignored. In these cases, the sample is supported by floating it or submerging it in a bath of hot oil at a controlled temperature. If it is floated, extra care must be taken to ensure temperature uniformity, and if it is submerged, the densities of the melt and the oil must be closely matched.

Another problem that arises is the change in shape of the sample as the deformation proceeds. For example, in uniaxial elongation, a cylindrical specimen becomes longer and thinner so that the uniformity of cross-section is increasingly difficult to control. If the stretching tractions are provided at the ends of the specimen, the maximum strain achievable will be limited by the length of the oil bath. Also, the force required to maintain a constant stress or strain rate will decrease markedly during the experiment, making it difficult to select an appropriate transducer.

A third problem arises as a result of the need to apply a tensile traction to the ends or edges of the sample. This requires the transmission of the tensile stress across a liquid-solid interface in such a way that the "clamping" fixture does not cause a significant nonuniformity of the deformation in the neighborhood of the end or edge.

The unique problems that arise in the measurement of extensional viscosities make it especially important to demonstrate the validity of any new technique. However, this is made difficult by the fact that Newtonian fluids cannot be used as calibration standards. Certainly, as a minimum demonstration that the desired property is being measured, the extensional viscosity at very low extension rates should be compared with the zero shear viscosity to see if Equation 2-27 is satisfied. For this purpose, it is best to use a resin with a Newtonian shear rate range easily accessible in a cone plate viscometer. A better verification can be achieved by comparing the results from two instruments of different design.

Methods for generating extensional flows in the laboratory have been described previously by Cogswell [221], Dealy [222] and Petrie [21].

6.1 UNIAXIAL EXTENSIONAL FLOW

This flow was defined in Section 2.3.3, where it was shown that the velocity distribution could be expressed as follows in cylindrical coordinates.

$$v_z = \dot{\epsilon} z \quad (\dot{\epsilon} > 0) \tag{6-1a}$$
$$v_r = -\tfrac{1}{2} \dot{\epsilon} r \tag{6-1b}$$

The existence of an axis of symmetry in the laboratory frame of reference does not imply that the deformation is not homogeneous from the point of view of fluid elements. Moreover, the cross-sectional shape of the sample is irrelevant as long as only normal tractions are applied to opposite ends and the cross-sectional area is uniform. For practical reasons, however, circular cylindrical specimens are usually used.

6.1.1 Basic Equations for Simple Extension

From Equation 6-1a, we see that the velocity in the direction of stretch is proportional to the distance from the plane at which $z = 0$, where the material is constrained from motion in the stretching direction. Thus, if the specimen being stretched has a total length L, the velocity of the moving end is as follows.

$$V = \dot{\epsilon} L \tag{6-2}$$

Since the velocity is simply dL/dt, the strain rate, $\dot{\epsilon}$, is obviously the rate of change of the Hencky strain, which was defined in Section 1.4.

$$\epsilon \equiv \ln(L/L_0) \tag{6-3}$$

Since we are assuming the material to be incompressible, the length and area at any instant are related as follows.

$$LA = \text{volume} = L_0 A_0 \tag{6-4}$$

Now we wish to establish a relationship between the rheologically meaningful normal stress difference, $(\tau_{zz} - \tau_{rr})$, and the tensile force, F, applied to the liquid specimen by the rheometer fixtures. We will define this force in such a way that it is the applied tensile traction in excess of the ambient normal force, $(-P_a A)$. Thus, the quantity, F, is related to the total tensile stress in the specimen by Equation 6-5.

$$\sigma_{zz} = F/A - P_a \tag{6-5}$$

If surface tension and inertia are neglected, the radial stress is uniform and equal to the ambient normal stress.

$$\sigma_{rr} = -P_a \tag{6-6}$$

Thus, the rheologically meaningful normal stress difference is

$$\sigma_e \equiv \tau_{zz} - \tau_{rr} = \sigma_{zz} - \sigma_{rr} = F/A \tag{6-7}$$

and, by use of Equation 6-4, this can be written in terms of measurable quantities as Equation 6-8.

$$\sigma_e = \frac{FL}{A_0 L_0} \tag{6-8}$$

The most common type of stretching experiment is the stress growth test, in which the strain rate is suddenly increased from zero to a constant value, $\dot{\epsilon}_0$, at time $t = 0$.

$$\left. \begin{array}{ll} \dot{\epsilon} = 0 & t < 0 \\ \dot{\epsilon} = \dot{\epsilon}_0 & t > 0 \end{array} \right\} \tag{6-9}$$

If this precise strain history is generated, then Equation 6-2 can be integrated to yield

$$L(t) = L_0 \exp(\dot{\epsilon}_0 t). \tag{6-10}$$

This implies that the area decreases exponentially,

$$A(t) = A_0 \exp(-\dot{\epsilon}_0 t) \tag{6-11}$$

and that the velocity of the moving end of the specimen increases exponentially.

$$V(t) = \dot{\epsilon}_0 L_0 \exp(\dot{\epsilon}_0 t) \tag{6-12}$$

The stress growth function is defined as follows.

$$\eta_T^+(t, \dot{\epsilon}_0) \equiv \sigma_e(t)/\dot{\epsilon}_0 \tag{6-13}$$

For the "ideal" stress growth experiment defined by Equation 6-9, this can be determined from the experimentally measured quantity, $F(t)$, by the use of Equations 6-8 and 6-10.

$$\eta_T^+(t, \dot{\epsilon}_0) = \frac{F(t) \exp(\dot{\epsilon}_0 t)}{\dot{\epsilon}_0 A_0} \tag{6-14}$$

In a creep experiment, the stress, F/A, is constant, and the strain is determined as a function of time.

$$\begin{aligned} F &= 0 \qquad t < 0 \\ F &= \sigma_0 A \quad t > 0 \end{aligned} \qquad (6\text{-}15)$$

The creep compliance in extension is then

$$D(t, \sigma_0) = \epsilon(t)/\sigma_0. \qquad (6\text{-}16)$$

If at a certain time, t, in the experiment the stress is suddenly removed, the specimen will undergo elastic recoil from length $L(t)$ to length $L_r(t)$, and the recoverable strain, ϵ_r, is given by Equation 6-17.

$$\epsilon_r(t) = \ln\left[\frac{L(t)}{L_r(t)}\right] \qquad (6\text{-}17)$$

Note that the time, t, is the time from the beginning of the creep test to the instant at which the stress is removed, and not the time after the stress is removed. If the creep test has proceeded to the stage where the strain is linear with time, then the recoverable strain should be independent of t.

6.1.2 Constant Stress Extensiometers

Pioneer work on the constant stress elongation of polymer melts was carried out by Cogswell [223], who used this technique to study a number of polymers [224]. In Cogswell's rheometer, a rod-shaped specimen is held at each end by a water-cooled collar, one fixed and the other mounted on a wheeled trolley. The little trolley has fastened to it a light chain that passes under a pulley at the end of the thermostatted oil bath, up out of the bath, and is fastened at its other end to a larger pulley. This pulley is mounted on the same axis as a specially designed cam that provides a moment arm for the action of a suspended weight, this moment arm varying with angle of rotation in such a way that the force exerted on the sample decreases as the sample extends so that the stress remains constant.

Design equations for such a cam have been presented by Dealy et al. [225], who also studied the limitations placed on the use of this equipment by hydrodynamic drag and bearing friction. As had been pointed out by Cogswell [223], these factors limit the use of this apparatus to the study of materials having extensional viscosities greater than 10^5 Pa·s.

A considerably more elaborate constant stress extensiometer was developed by Vinogradov et al. [226]. In this apparatus, the clamps float so that hydrodynamic drag is reduced, and the strain is measured by a photoelectric follower which introduces no friction.

Of course, as the experiment proceeds the force becomes smaller and smaller, and frictional forces ultimately become significant. Münstedt [227] has developed a constant stress extensiometer that reduces the instrument friction to a minimum. The sample is mounted vertically in an oil bath, thus eliminating one pulley. The small moving clamp is fastened to the sample by means of an adhesive and makes a negligible contribution to hydrodynamic drag. The remaining pulley is mounted on an air bearing. This device seems to yield reliable data for branched polyethylenes [228, 229].

6.1.3 Constant Velocity at a Fixed Point

We see from Equation 6-1a that the desired deformation can be achieved, as long as compressibility and inertia can be neglected, by providing for a specimen to be fixed in position at one end ($z = 0$) and to have a constant velocity at any fixed point, z_1. If this velocity is suddenly increased from zero to the constant value, $V(z_1)$, at time zero, and the force required to maintain this velocity is measured as a function of time, then the stress growth function can be determined. The strain rate is given by

$$\dot{\epsilon}_0 = v(z_1)/z_1. \qquad (6\text{-}18)$$

Meissner [230] developed a constant strain rate rheometer based on a variation of this concept. In this device, constant velocities, v_1 and v_2, in opposite directions, are maintained at two locations a distance Z apart. The strain rate is then given by

$$\dot{\epsilon}_0 = \frac{v_1 + v_2}{Z}. \qquad (6\text{-}19)$$

This is accomplished by means of two "rotary clamps," each consisting of a set of wheels with small gear teeth, which are driven by a synchronous motor and rotate in opposite directions with the sample gripped between them. Meissner derived an equation giving the effective length, Z, of the stretched portion of the sample. One set of gears is mounted, along with its drive motor, on a vertical leaf spring fixed at its upper end. A sensitive displacement transducer (LVDT) is used to produce a signal proportional to the deflection of the spring and thus to the tensile force in the melt. An important advantage of the use of rotary clamps is that the maximum strain is not limited by the length of the rheometer. The rotary clamps are submerged in an oil bath along with small scissors to allow the cutting of short lengths of stretched material to be used to determine the recoverable strain, $\epsilon_r(t)$, and the uniformity of the stretched specimen.

Meissner [231] later showed how this rheometer can be modified to carry out a constant stress creep test or a deformation with any desired strain history, $\dot{\epsilon}(t)$, in addition to the special one defined by Equation 6-9. This is accomplished by replacing the synchronous motor driving one rotary clamp by a servomotor and operating the remaining synchronous motor (the one mounted on the leaf spring) at a speed just sufficient to maintain effective clamping. For operation at constant stress, a signal proportional to stress is generated and used as the input to a control system for the servomoter. This signal is generated by an analog circuit which operates on voltages proportional to the stress and the angular displacement, ϕ, of the rotary clamp wheels. By modifying this circuit, arbitrarily defined stress and strain patterns can be generated.

Meissner [232] used the rotary clamp rheometer to study several low density polyethylenes, but was not able to achieve sufficiently large strains to reach steady state except at low strain rates in the range of linear viscoelasticity. Laun and Münstedt [228, 229] modified the force measuring system of this rheometer to reach Hencky strains as large as 6, at which the elongation ratio is about 400. In a comparative study of several polyethylenes [229, 233] using both this rheometer and the creepmeter mentioned in the previous section, they were able to determine the extensional viscosity function, $\eta_T(\epsilon)$, over a range of strain rates from 10^{-5} to 10 s^{-1}. To reach the lowest decade of this range, one of the rotary clamps was replaced by a piece of metal to which the sample was fixed by an adhesive. This piece of metal was attached directly to a leaf spring, the deflection of which was used to measure the tensile force.

Raible *et al.* [234], noting that the maximum strain in this type of test is limited by the uniformity of the temperature of the melt, designed a rotary-clamp rheometer in which it is possible to keep temperature variations below 0.1°C at 150°C. They were thus able to generate deformations up to a Hencky strain of 7. This implies an extension ratio (L/L_0) of over 1,000!

Another design making use of the rotary clamp idea is that of Cotton and Thiele [235]. In their rheometer, a long strand of raw rubber is looped over a set of pulleys supported from a load cell, and both ends are passed through a single set of gripping rollers driven by a constant speed motor. The measured force is then the sum of the tensions in the two segments of the strand.

Macosko and Lorntson [236] used a commercial rotational rheometer, the Rheometrics Mechanical Spectrometer,* to generate uniaxial extension in high density polyethylenes at 180°C. Heating was provided by a forced convection air oven so that the sample had to support its own weight. The ends of the sample were attached to slotted vertical spindles separated by a distance, Z, limited to rather small values by the design of the rheometer. One of the spindles (the "winder") was rotated at a constant speed, Ω, while the lateral force, F, exerted by the melt on the second spindle, was measured. The winder was raised during the experiment to prevent the liquid fiber from wrapping on itself. The extension rate is given by

$$\dot{\epsilon} = \Omega R/Z \qquad (6\text{-}20)$$

where R is the radius of the winder spindle. The stress growth function was computed by use of Equation 6-14. Because Z is small, end effects are a problem, especially at large strains, and Macosko and Lorntson were unable to extend their stress growth tests to strains sufficiently large for the stress to reach a steady state value. Everage and Ballman [237] used the same technique to study the extension of polystyrene at 155°C and were also unable to reach a steady state.

Connelly *et al.* [238] explored the limitations on the use of this

---

*Described in Section 10.3.

instrument for extensional flow studies of melts. They found that when the product of the strain rate and the fluid's longest relaxation time was larger than 1, the response was largely elastic and the deformation was uniform. When this product was less than 1, however, the behavior was largely viscous, and uniform strain was observed only at very small strains. Thus, the limitations to small values of R and Z imposed by the design of the Mechanical Spectrometer limit its usefulness in the study of extensional flows.

Ide and White [239, 240] built a constant strain-rate extensiometer similar in concept to the Mechanical Spectrometer, but in which the winder radius and stretched length are much larger than is possible in that instrument. In their rheometer, the sample is floated on an oil bath with one end clamped to an Instron load cell and the other end fastened to the rim of a relatively large wheel driven by a constant speed motor. They studied a wide range of polymers, including high and low density polyethylenes, polypropylene, polymethylmethacrylate and polystyrene. In many cases, they were unable to reach a state of steady stress before the sample broke.

6.1.4 Programming of Velocity or Length

Another approach to the problem of generating a constant strain rate is to provide for the linear translation of one of two end clamps such that the velocity of the moving clamp follows the function given by Equation 6-12. Most of the studies using this technique have been carried out by use of a standard tensile testing machine with the velocity signal provided by a variable function generator or an analog computer. These include the work of Ballman [241], Stevenson [242] and Shaw [243]. Shaw used an interesting technique to handle the clamping and end-effects problems. His samples were molded, circular rings, which were looped over one fixed and one moving hook. To start the experiment, the ring was first stretched to form two parallel fibers connected together at their ends.

Vinogradov et al. [244] built a unique rheometer based on a preprogrammed velocity function. In their instrument, the moving clamp was driven by a worm-gear and rack mechanism. The velocity signal was programmed by means of a photoelectric servo mechanism following a curve drawn on a piece of paper.

In assessing the validity of results obtained by use of a velocity preprogrammed to follow the function of Equation 6-12, it is necessary to note that the velocity cannot suddenly rise from zero to the desired finite value at $t = 0$, $V(0) = \dot{\epsilon}_0 L_0$, because of instrument inertia. Thus, at the beginning of an experiment, there is a period during which the velocity is less than the value given by Equation 6-12. The length, on the other hand, has the correct initial value, L_0, given by Equation 6-10 for $t = 0$, but since the length is equal to the integral of the velocity, and since the velocity is initially lower than the desired value given by Equation 6-12, the length will fall behind the theoretical value given by Equation 6-10, and it will catch up with this exponential function only asymptotically at very long times. Thus, the strain rate, as given by Equation 6-2, is zero at $t = 0$, rises to a value higher than the setpoint value, $\dot{\epsilon}$, and then gradually approaches $\dot{\epsilon}_0$ asymptotically. This problem can be eliminated by use of an appropriate feedback control system, as described in Section 6.1.5.

Agrawal *et al.* [245] avoided this problem by preprogramming $L(t)$ rather than $V(t)$. This was accomplished by the use of a constant speed motor geared to produce translational motion of a "linear cam," which, by means of a mechanical linkage, controlled the displacement of the moving clamp. By altering the shape of the "linear cam," we see from Equation 6-3 that any arbitrary strain history can thus be generated.

6.1.5 Feedback Control of Velocity or Length

If a sample is stretched by means of the rectilinear displacement of a moving clamp, and if a voltage proportional to sample length, $L(t)$, can be generated, this voltage can be used as the input signal for a control system for either a tachometer feedback motor, so that velocity is controlled, or for a servomotor, so that extension is controlled. If the velocity of a moving clamp is maintained proportional to L, then the result is a constant-strain-rate test.

$$V(L) = \dot{\epsilon}_0 L \qquad (6\text{-}21)$$

Thus, the adherence to the desired strain history does not depend on Equation 6-10 and is limited only by the reliability of the control system. By use of more elaborate circuits, such a system can be used to generate a constant stress test or one in which the strain rate is any prescribed function of time, $\dot{\epsilon}(t)$.

Rhi-Sausi and Dealy [246] designed an extensiometer based on this principle. The moving clamp is suspended from a ball nut moving along a long lead screw driven by a tachometer feedback motor,* the driving voltage for which is proportional to the sample length as measured by a multiturn potentiometer.

Münstedt [247] has presented the design of an extensional rheometer that combines many of the attractive features of his creepmeter [227] with the flexibility and precision of feedback control.** The basic features of the Münstedt extensiometer are shown in Figure 6-1. The fixed clamp consists of a small piece of sheet metal hooked to a load cell at the bottom of an oil-filled, cylindrical, jacketed glass vessel. The load cell is mounted on a long glass rod extending upwards and out of the bath to a support frame. The use of a glass rod minimizes heat transfer out of the bath due to conduction. The second clamp is hooked to the end of a flexible tape that extends upwards and winds onto a wheel

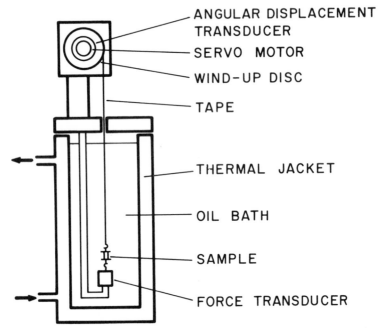

Figure 6-1. Münstedt's extensiometer [247].

---

*Electrocraft, Hopkins, Minn.
**Two commercial rheometers based on this design are described in Section 11.2.

mounted above the bath on the shaft of a DC servomotor. An angular displacement transducer mounted on this shaft produces a signal proportional to sample extension. This rheometer can be used for constant strain rate and constant stress tests as well as programmed strain tests such as superposed oscillatory and steady extension.

Laun and Münstedt [229] examined carefully the possible effect of interfacial tension on the calculation of the rheologically meaningful stress difference, σ_e, by the use of Equation 6-7. If interfacial tension is taken into account, this equation must be modified as follows,

$$\frac{F}{A} = \sigma_e + \frac{\alpha}{R}\left(1 - \frac{2R}{L}\right) \qquad (6\text{-}22)$$

where α is the interfacial tension. This equation is written for the case in which a specimen of initial length, L_0, is stretched by applying traction to its ends. In the case of the use of rotary clamps, the second term in the brackets vanishes. Laun and Münstedt [229] showed that for a low density polyethylene at 150°C the interfacial tension makes a significant contribution to the total tensile force only at strain rates below 10^{-4} s^{-1} and becomes the dominant factor at strain rates below 10^{-5} s^{-1}.

6.2 SYMMETRICAL BIAXIAL EXTENSION

This deformation, first mentioned in Section 2.3.4, has the following velocity distribution in cylindrical coordinates.

$$v_r = \dot{\epsilon}_b r \qquad (6\text{-}23\text{a})$$
$$v_z = -2\dot{\epsilon}_b z \qquad (6\text{-}23\text{b})$$

There is one rheologically meaningful normal stress difference, and we will call this σ_b.

$$\sigma_b \equiv \tau_{rr} - \tau_{zz} \qquad (6\text{-}24)$$

In an experiment in which the strain rate, $\dot{\epsilon}_b$, increases from zero to some constant value at time $t = 0$, one can determine the biaxial stress growth function defined as follows.

$$\eta_b^+(t, \dot{\epsilon}_b) \equiv \sigma_b(t)/\dot{\epsilon}_b \qquad (6\text{-}25)$$

The generation of this deformation has proven considerably more difficult than that of uniaxial extension.

Meissner et al. [248] have proposed the use of the rotary clamp concept to generate biaxial extension. Eight pairs of rotary clamps are arranged in a circle, and eight pairs of automated scissors cut the sheet of molten polymer between the clamps at frequent intervals. This cutting produces eight strips of polymer that can be wound up on rollers. The cutting introduces periodic fluctuations in the measured tensile force, but the authors report that meaningful values of time-dependent stress can be determined from the experimental data. To date, this technique has been used only with polyisobutylene at room temperature, but work is continuing to extend it to the study of molten thermoplastics.

Stevenson [249] has suggested the use of "lubricated squeeze flow" for the generation of biaxial extension. If a melt is squeezed between two (unlubricated) parallel flat plates, a rather complex flow results that is neither a viscometric flow nor an extensional flow.* However, if the two plates are coated with a layer of relatively inviscid liquid that prevents contact between the melt and the wall, then the flow defined by Equation 6-23 can be generated. If the strain rate, $\dot{\epsilon}_b$, is to be constant, we find from Equation 6-23b that the velocity of the moving plate, V, must decrease in magnitude as the plates approach each other.

$$V = -2\dot{\epsilon}_b h \qquad (6\text{-}26)$$

If the strain rate could actually be increased from zero to a constant value, $\dot{\epsilon}_b^0$, instantaneously, then the plate spacing would follow an exponential decay.

$$h = h_0 \exp(-2\dot{\epsilon}_b^0 t) \qquad (6\text{-}27)$$

The compressive force, F, in excess of $A \cdot P_a$ required to generate the lubricated squeeze flow, is related to the normal stress difference, σ_b, as follows,

$$\sigma_b \equiv \tau_{rr} - \tau_{zz} = F/A \qquad (6\text{-}28)$$

---

*The use of this flow to characterize melts is discussed in Chapter 7.

where it has been assumed that inertia and surface tension are negligible and that the edge of the sample is exposed to the ambient pressure, P_a. If the material is incompressible, then the area, A, will vary with time,

$$A(t) = A_0 h_0 / h(t) \qquad (6\text{-}29)$$

and σ_b is related to the measurable quantities, h and F, according to Equation 6-30.

$$\sigma_b = F h / A_0 h_0 \qquad (6\text{-}30)$$

The biaxial strain, ϵ_b, is

$$\epsilon_b = \tfrac{1}{2} \ln(h_0 / h). \qquad (6\text{-}31)$$

Chatraei and Macosko [250] used lubricated squeeze flow as the basis of a creepmeter, which they used to measure the biaxial extensional viscosity of a polydimethylsiloxane gum. The initial sample diameter was the same as that of the rheometer disks, so that as the squeezing progressed, the displaced fluid flowed out around the edge, leaving the effective sample area unchanged. In this way, the application of a constant force, by means of dead weight loading, provided a constant stress. They compared the use of petrolatum and a silicone oil as lubricants.

6.2.1 Inflation of a Circular Sheet

The only technique that has been used with success to study the behavior of molten polymers in flows approximating axisymmetric biaxial extension is sheet inflation. A sheet of the material to be studied is clamped between two plates, both of which have circular concentric holes cut in them. An inflation medium, often a gas, is introduced under pressure to a chamber mounted on one side of the plate. The pressure difference, ΔP, between the two sides of the sheet and the deformation of the sheet are monitored as functions of time, providing data from which a material function is determined.

We will consider for the moment an idealized sheet inflation in which the deformation is assumed to be uniform over the entire sheet so that a spherical shell of uniform thickness is formed as shown in Figure 6-2.

The tangential stress minus the stress component normal to the surface is associated with the net stretching stress, σ_b, of biaxial extension. Classical shell analysis yields the following relationship,

$$\sigma_b = R \cdot \Delta P/(2\delta) \tag{6-32}$$

where R is the radius of curvature of the shell and δ the sheet thickness.

The stretch ratio, λ, is associated with the change in surface distance (arc length) from the pole to the clamped edge of the sheet so that

$$\lambda = 2\alpha R/D \tag{6-33}$$

where D is the diameter of the hole in the clamping plate through which the sheet is inflating, and α is defined as shown in Figure 6-2. The definition of the sine function can be used to show that

$$\lambda = \alpha/\sin\alpha.$$

The strain, ϵ_b, is then given by

$$\epsilon_b = \ln\lambda = \ln\left(\frac{2\alpha R}{D}\right) = \ln\left(\frac{\alpha}{\sin\alpha}\right). \tag{6-34}$$

Since α and R are often not as easily measured as the sheet height, h, the following geometric and trigonometric relationships are useful for data analysis.

$$h = R(1 - \cos\alpha) \tag{6-35}$$
$$D/2R = \sin\alpha \tag{6-36}$$

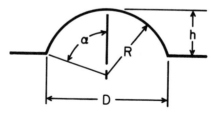

Figure 6-2. Quantities used to describe uniform sheet inflation.

Eliminating α between these two expressions, we have

$$R = D^2/8h + h/2. \tag{6-37}$$

Noting that a hemispherical shell occurs when $R = D/2(\alpha = \pi/2)$, this condition can be found to be associated with a Hencky strain of 0.45.

The surface area of a spherical segment is $(2\pi Rh)$, so that if the material is assumed to be incompressible, the thickness is given by

$$\delta = \delta_0 \left(\frac{D^2}{8Rh} \right). \tag{6-38}$$

This quantity is needed to calculate the stress according to Equation 6-32. Thus, knowledge of ΔP together with any one of the three quantities, α, R or h, as a function of time, with $\dot{\epsilon}_b$ maintained constant, is sufficient for the determination of the stress growth function, $\eta_b^+(t,\dot{\epsilon}_b)$.

In fact, the uniform sheet inflation described above is not kinematically equivalent to the deformation defined by Equation 6-23, except when the radius, R, is very large. For example, we note that the area of a flat circular sheet is not related to its diameter in the same way that the surface area of a spherical segment is related to the surface distance from the pole to the circle defining its base. Furthermore, the clamping plates restrain the displacement of fluid particles adjacent to them so that uniform stretching over the entire sheet is not possible.

These facts would seem to preclude the use of the sheet inflation technique to study biaxial extension. However, observations of inflated sheets of rubber [251, 252] indicate that there is a region near the pole in which the deformation is both uniform and a satisfactory approximation of the flow defined by Equation 6-23, and studies involving certain viscoelastic melts [253, 254] have indicated that this may also be true for these materials.

In this case, measurements must be made on a small segment of the spherical shell in the neighborhood of the pole. Such a segment is shown in Figure 6-3. The strain, based on stretching in the longitudinal direction, is given by

$$\epsilon_b \text{ (long.)} = \ln \left(\frac{2\beta r}{d_0} \right) = \ln \left(\frac{d\beta}{d_0 \sin\beta} \right) \tag{6-39}$$

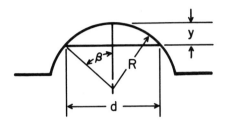

Figure 6-3. Sheet inflation; the region near the pole.

where all the quantities are defined in Figure 6-3 except d_0, which is the diameter prior to inflation, of the circle forming the base of the segment. The strain in the latitudinal direction is

$$\epsilon_b \text{ (lat.)} = \ln (d/d_0). \tag{6-40}$$

If β is small, these two strains become practically equal as should be the case in "ideal" biaxial extension.

Equations 6-32 and 6-35 to 6-37 can be modified for use with the segment by replacing R, α, h and D by r, β, y and d, respectively. The sheet thickness is now given by

$$\delta = \delta_0 \left(\frac{d_0^2}{4yd} \right) \sin \beta$$

but if β is small, this is approximately equivalent to

$$\delta = \delta_0 (d_0/d)^2. \tag{6-41}$$

Sheet inflation has been used to study the biaxial stretching of thermoplastics in the rubbery state by several groups [252, 255, 256]. De Vries and Bonnebat [252] used an elaborate apparatus to generate a constant strain rate flow. A caliper-like device with its two tips "welded" ultrasonically to the sample surface near the pole acts as a strain transducer, and the signal generated is used to operate a feedback control system regulating the flow of air into the inflating sheet.

The direct measurement of strain is not so straightforward when the material studied is a melt, and photographic records are usually used for this purpose. Photos of the expanding sheet were used, for example, by Denson and Gallo [257] to study polyisobutylene stretching at room temperature. They analyzed their data by use of Equations 6-32 and 6-34, assuming the inflated sheet to be uniformly stretched.

Joye *et al.* [253] noted that the use of the average values of σ_b and ϵ_b, calculated assuming uniform strain, yields only approximate results. In the analysis of their data, they focused attention on the region near the pole and compared several methods for estimating the sheet thickness in this region. They found that the strain rate near the pole was greater than the average value calculated from Equation 6-34, but that the two values tended to be proportional to each other.

Whereas the experiments of Joye *et al.* [253] were carried out at approximately constant stress, Maerker and Schowalter [254] regulated the flow rate of the inflating gas to approximate a constant strain rate. These authors studied materials with lower viscosities than any used previously and concluded that the air inflation method is limited to materials having values of η_0 above 10^5 Pa·s.

The sheet inflation devices described above are designed for use at ambient temperature with a gas as the inflation medium. Thus, operation at elevated temperatures was not possible, and only very high viscosity materials could be studied, as there was no support for the sample prior to the inflation process. Furthermore, the direct control of the inflation rate was not possible because of the compressibility of the gas.

The use of a liquid as the inflation medium makes possible the support of the sample prior to inflation and the direct control of the inflation rate. Denson and Hylton [258] used a liquid inflation medium, although their apparatus was designed to characterize materials at temperatures typical of thermoforming processes. Thus, the materials they studied were in a rubbery state and exhibited very large viscosities.

Dealy [259] described some years ago an apparatus for the study of thermoplastics at typical melt processing temperatures. A premolded, disk-shaped sample is placed between two water-cooled rings which act as a clamp, and the clamp assembly is mounted atop a chamber containing silicone oil. A second oil-filled chamber is mounted above the clamp for temperature control and to counterbalance the hydrostatic gradient across the sample resulting from the inflation medium. The oil has a density matching that of the melt to prevent sagging or floating.

A cylinder fitted with a piston is mounted in the side wall of the lower chamber so that the movement of the piston displaces oil into the lower chamber driving the inflation of the sheet. The motion is driven by a linear actuator coupled to a servomotor.

Rhi-Sausi and Dealy [260] used an extensiometer of the type described above, and reported great difficulty in obtaining reliable results due primarily to temperature nonuniformities and the unavoidable trapping of air bubbles under the inflating sheet. They designed an improved device in which the sheet is mounted vertically instead of horizontally to eliminate the trapping of air. Also, they designed a clamp which does not require water cooling, so that this source of temperature nonuniformity was eliminated. They reported that the improved rheometer gives reliable results, but that the maximum attainable strain is limited to a value in the range of 1.3 to 1.7, so that the entire stress growth curve cannot be obtained except at very low strain rates.

6.3 PLANAR EXTENSION

This flow, also called "pure shear," was defined in Section 2.3.5. Denson and Crady [261] and Denson and Hylton [258] have approximated this flow by use of a sheet inflation technique with the circular hole replaced by a rectangular slit with a large aspect ratio. The rim effects are not a severe problem in this case, and the assumption of uniform cylindrical inflation is found to be valid for practical purposes. In this case, the basic equations for calculating the measurable stress difference and the strain are as follows,

$$\tau_{11} - \tau_{22} = \frac{\Delta P \cdot R}{h} \tag{6-42}$$

$$\epsilon = \ln\left(\frac{2\alpha R}{D}\right) \tag{6-43}$$

where the quantities are as defined in Figure 6-2. Of course, the inflated sheet in this case is cylindrical rather than spherical.

6.4 GENERAL EXTENSIONAL FLOWS

Chung and Stevenson [262] have proposed a general scheme of nomenclature and classification for the entire family of flows whose kinematics are described by Equation 2-15, and Stevenson et al. [263] have proposed an experiment which could, in principle, be used to determine the material functions that are associated with them. This is based on the simultaneous extension and inflation of a tubular sample.

Chapter 7
Complex Flows Used To
Characterize Melts

A number of flows have been used to characterize viscous liquids that do not fit into any of the categories so far considered. They are neither viscometric nor flows with constant stretch history. As a result, their usefulness in the determination of well-defined material functions is quite limited. Some can yield reliable information about low shear rate viscosity, and in other cases it has been claimed that there is a correlation between the measured quantity and some material function. In any event, several of these complex flows are widely used in the plastics industry, and it is important to understand what kinds of information they can and cannot provide.

7.1 THE FALLING SPHERE RHEOMETER

As has been noted in Chapter 2, it is often difficult to obtain data at a sufficiently low shear rate to determine the limiting Newtonian viscosity, η_0. Foltz *et al.* [264] have developed a rheometer especially suited to the measurement of η_0 over a range of pressures. The sample holder is a steel cylinder fitted with a piston that acts from below to provide the desired pressure level. A metal sphere, slightly more dense than the melt, is placed between two premolded lengths of resin, which are loaded into the top of the cylinder with resin pellets filling the remainder of the space. The geometry of these premolded pieces governs the initial position of the sphere. A set of cylinders, each at a different pressure, is placed in an oven for 20 to 48 hours after which they are cooled, and the location of the sphere is determined. The viscosity is calculated from Stokes' law, making the appropriate corrections for wall effects. An obvious source of error in this test is the possibility of polymer degradation during the long heating period.

7.2 ROD CLIMBING

If a vertical rod is rotated while partially submerged in a pool of viscoelastic liquid, the liquid will rise up around the rod and assume a steady state free surface shape that is characteristic of the material's rheological nature. This phenomenon is often referred to as the Weissenberg effect. Experimental studies of polymer solutions have shown that the streamlines are not circular and that there are always secondary circulations. However, the magnitude of the secondary component of the velocity is often very small, and it has been suggested—for example, by Hoffman and Gottenburg [265]—that the viscometric functions can be estimated from measurements of the free surface flow.

Dealy and Vu [266] have discussed the possibility of using this flow to characterize molten polymers. However, they found that for many commercial resins, the time required for the free surface to reach its steady state configuration is in the range of 1 to 2 hours, so that degradation becomes a serious problem. They suggested an empirical criterion for estimating when this problem is likely to arise. Specifically, they found that the starting transient is of the order of a few seconds when $(N_1/0.21\ S\eta)$ is much less than 1, where S is the rotational speed in RPM, and N_1 and η are evaluated at $\dot{\gamma} = 0.21\ S$. However, even in this case, the characterization is an empirical one, and there is no generally reliable relationship between free surface shape and well-defined material functions.

7.3 TORQUE RHEOMETER FLOW

A popular type of testing equipment in the plastics industry is the so-called "torque rheometer," commercial versions of which are described in Section 11.4. We will discuss here only the extent to which rheological material functions can be inferred from torque rheometer data.

A torque rheometer consists of a horizontally mounted, heavy-duty motor drive together with a torque sensor. A variety of devices are available for attachment to the drive shaft, including mixing heads and small, single-screw extruders. One of the most used devices is the "roller type" mixing head, in which two rotors of rather complex shape rotate at different speeds within adjoining circular cavities. The flow is neither uniform nor controllable, and we should thus expect that the details of the flow pattern will depend on the rheological properties of the fluid under study. Also, since the strain history is not constant, elasticity is likely to play an overt role in the material response. Since the flow chamber is not thin, temperature nonuniformities due to viscous heating will be significant.

The complexity of the flow in one type of mixing head, the Banbury, has been demonstrated by Freakley and Idris [267] by use of a transparent mixing chamber and a transparent rubber.

Nevertheless, there have been attempts to relate torque versus RPM data from torque rheometers to the shear viscosity function. Goodrich and Porter [268] assumed that the two shearing zones could be considered in some sense as equivalent to two concentric-cylinder rheometers operating at different shear rates. The appropriate apparatus constants were determined experimentally by the use of high-viscosity Newtonian fluids.

Blyler and Daane [269] and Lee and Purdon [270] attempted to derive relationships that would be valid for power law fluids. Based on different sets of assumptions, both pairs of authors derived the following equation,

$$M = F(n) k \Omega^n \qquad (7\text{-}1)$$

where $F(n)$ is a function of the power law index and, of course, the mixer geometry. This result implies that the shear rate distribution in the mixer depends only on the index, n. While it is possible that this may be valid for certain materials over a certain range of speeds, it cannot be expected to have universal validity.

Rogers [271] has proposed a method for interpreting data obtained from an extruder head in terms of viscosity. The flow is considered to be equivalent to that in a concentric cylinder rheometer. He also examines the flow in a cylindrical die attached to the extruder and interprets the ΔP versus Q data using the techniques described in section 4.3. Yasenchak [272] and Mentovay and Yasenchak [273] also used data on an extruder-fed die to determine the viscosity, although the latter authors used an incorrect definition of viscosity.

In connection with the use of extruders for melt testing, the reader is referred to the description of the Revesz screw-tip torque sensor given at the end of Section 5.1.

7.4 EXTRUDATE SWELL

At the exit of a capillary, slit or die of any shape an emerging stream of polymeric liquid undergoes a substantial increase in cross-sectional area from that of the channel in which it has been flowing. This phenomenon is usually called die swell or extrudate swell. If the channel is noncircular, then the extrudate will also undergo a change of shape. Extrudate swell can occur even in the case of Newtonian liquids at low Reynolds numbers [274, 275, 276], but the effects are much more pronounced in the case of viscoelastic liquids. Of course, changes in the shape and size of an extrudate can be very troublesome for manufacturers of all kinds of extruded profiles, but they have also been of considerable interest to melt rheologists.

In the case of capillary extrudate, there is no change of shape so that the swell can be quantitatively described in terms of the extrudate diameter, and this is the type of extrudate swell that has received the most attention from melt rheologists. Figure 7-1 shows a sketch of the type of behavior exhibited by many polymeric liquids. There is a rather rapid swelling, which occurs almost immediately upon efflux, as can be seen here, but in a melt the swelling can continue for quite a long time. In addition, the swell is found to depend on the wall shear rate in the die and on the L/D ratio.

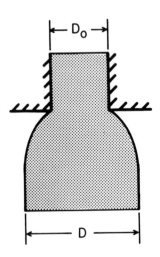

Figure 7-1. Capillary extrudate swell.

The "die swell ratio," B, is defined as the ratio of the diameter of the extrudate, D, divided by that of the die or capillary, D_o.

$$B \equiv D/D_0 \qquad\qquad (7\text{-}2)$$

As was mentioned above, for a given die entry shape, this ratio is a function of the wall shear rate, $\dot{\gamma}_w$, the L/D ratio, and the time elapsing after exiting the die.

$$B = f(\dot{\gamma}_w, L/D, t)$$

Han [6] has summarized the findings of a number of experimental studies of the swelling of molten polymers. The swell ratio decreases as L/D increases, approaching a constant value at values of L/D often around 20. The dependency of B of L/D has been interpreted by Chapoy [277] in terms of capillary residence time, but Bagley *et al.* [278] have noted that it seems to be the total amount of shear strain occurring in the capillary that governs the approach of B to its limiting value for large L/D.

The quantity most often used to characterize melts is the equilibruim or ultimate swell, which is reached sometime after extrusion. This quantity should be independent of die entry shape and L/D when L/D is large.

Extrudate swell is not a viscometric flow, and the swell ratio cannot be related to the viscometric or linear viscoelasticity functions by any general, rigorous derivation. Nevertheless, because of the relative simplicity of the experimental technique, a number of equations have been proposed to relate the swell ratio to other material functions [279–284], and Vlachopoulos *et al.* [285] have compared the reliability of several of these. They are derived by making two types of assumptions. First, geometric assumptions are made to relate the unconstrained recoil that occurs in extrudate swell to the constrained recoil that occurs in creep recovery. Then, assumptions are made as to the general rheological behavior of the melt. For example, Equation 2-55 may be assumed to be valid even at high shear rates. These types of assumptions lead to a relationship between the recoverable shear, S_r, and the swell ratio, B. Finally, the use of an assumed relationship between recoverable shear and the viscometric functions, such as Equation 2-57, yields a relationship between the first normal stress difference and the swell ratio.

Tanner [275] carried out numerical calculations of the flow of a Newtonian fluid at the exit of a tube, and concluded that there is a basic fallacy in the geometric descriptions of extrudate swell used as bases for the equations mentioned above. Whipple and Hill [127] carried out an experimental study of the details of the flow of concentrated polymer solutions at the exit of a slit die and also concluded that the changes in shape which occur when fluid elements approach and exit the end of a die are "grossly different from those assumed to occur in elastic-like fluid theories."

At the present time, one can conclude only that a properly executed extrudate swell experiment yields an empirically defined material function that may be of use in classifying and comparing materials in specific applications, but that there is no reliable way to convert swell data into other material functions such as the recoverable shear or the first normal stress difference.

Several different techniques have been used to measure extrudate swell, although, for reasons to be explained shortly, these do not all yield equivalent values of B. In order to avoid sag, the extrudate is usually cooled immediately, and the average diameter of a solidified length, L, of extrudate is determined by measuring its mass, M. If ρ_s is the density of the sample at its cooled temperature, T_s, then the average extrudate diameter is given by

$$D(T_s) = \sqrt{\frac{4M}{\pi L \rho_s}}. \tag{7-3}$$

Pliskin [286] has described an instrument that can be used for the rapid measurement of extrudate swell. This device was designed for in-plant quality control testing of elastomers and consists of a piston-driven capillary extruder together with a pair of photocells mounted to permit the precise measurement of the time, t, required for the extrudate to move through a distance L. The velocity of the extrudate is (L/t) and the swell ratio is

$$\frac{D(T_s)}{D_0} = \left[\frac{V}{(L/t)}\right]^{1/2} = \left(\frac{Vt}{L}\right)^{1/2} \tag{7-4}$$

where V is the average velocity of flow in the die. Another device which makes a direct determination of swell during extrusion is the Monsanto Automatic Die Swell Detector, described in Section 9.4.3. It is based on the use of a sweeping laser beam to measure the extrudate profile.

In assessing the reliability of these methods, we note first that the density of the extrudate at its extrusion temperature, T_e, is different from the density of the cooled sample at T_s. In order to estimate the extrudate diameter at the temperature, T_s, we can assume isotropic thermal contraction to obtain the following relationship.

$$D(T_e) = D(T_s) \left[\rho(T_s)/\rho(\mathbf{T}_e)\right]^{1/3} \tag{7-5}$$

A more troublesome aspect of the methods mentioned above for determining B is that the value of time is not well-defined, and it is usually too short to allow the equilibrium swell to be attained. One approach to this problem is to reheat the cut lengths of extrudate in an oven to allow them to approach their equilibrium dimensions. Racin and Boque [282] extruded into a heated air chamber and took photographs of the extrudate to determine the swell. In their studies, they found sag to be an unimportant source of error. White and Roman [287] found this method to yield a higher value of B than the reheat method mentioned previously.

A more elaborate procedure, which allows one to observe swell as a function of time over long periods in the absence of sag, is to extrude the polymer into a bath containing a relatively low viscosity oil at the extrusion temperature, T_e. The density of the oil should be equal to or slightly less than that of the melt. This was the technique used by Utracki et al. [288] and by Dealy et al. [289]. They extruded into a series of oil-filled test tubes by cutting successive lengths of extrudate produced at a sequence of extrusion rates; i.e., at a sequence of wall shear rates. After waiting a sufficient length of time for the equilibrium swell to be attained, photographs of the diameters of the still molten extrudate samples were taken through a side window in the thermostatic bath containing the test tubes. By taking a series of photos of a single sample, Dealy and Garcia-Rejon [290] used this technique to study the time dependency of swell.

7.5 SQUEEZING BETWEEN PARALLEL PLATES

This class of parallel plate flows is of interest because of the ease with which they can be generated in the laboratory.* However, they are neither viscometric nor purely extensional flows, and the deformation is neither steady in time nor homogeneous in space. For these reasons, a rheometrical relationship can be derived only by assuming a form for the constitutive equation, and such relationships have been derived for the cases of Newtonian and power law liquids. In general, the utility of tests based on these flows is limited to flow regimes in which elasticity does not play an important role.

Two modes of operation have been used for parallel plate rheometers. In "squeeze film" flow, the sample initially has the same size as the disk-shaped parallel plates between which it is compressed. At the beginning of the test, a total force, F, is applied to the upper plate by means of weights, and the plate spacing, $h(t)$, is measured as a function of time. As the compression proceeds, material is extruded from the gap, and the total compression area remains constant and equal to the area of the disks. This mode of operation is illustrated in Figure 7-2. In a "parallel plate plastometer," also often called a "Williams plastometer," the sample is smaller in diameter than the plates, and it is the sample volume that remains constant rather than the compression area, as shown in Figure 7-3.

The theory of parallel plate rheometers is based on a number of simplifying assumptions. In particular, the inertia terms in the equation of motion are neglected, and the velocity distribution is assumed to depend

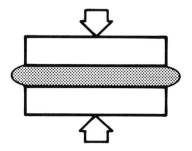

Figure 7-2. Squeeze film geometry.

---

*Commercial instruments which make use of this class of flows are described in Section 11.3.

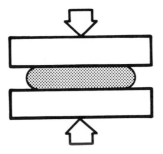

Figure 7-3. Parallel plate plastometer.

on the spatial variables in a certain way. We will consider first the theory of squeeze film (constant area) flow.

Bird et al. ([1], p. 19) give the derivation of the Stefan equation relating the force to the speed of the upper plate.

$$F = \frac{3\pi R^4 \eta(-dh/dt)}{8h^3} \tag{7-6}$$

Integrating this equation for a constant force, we obtain

$$\frac{1}{h^2} = \frac{1}{h_0^2} + \frac{16Ft}{3\pi R^4 \eta}. \tag{7-7}$$

However, the neglecting of inertia cannot be valid at the onset of the flow, so that if experimental values of $(1/h^2)$ are plotted versus time, the earliest points will not lie on a line intersecting the ordinate at $(1/h_0^2)$. If the data approach a straight line, this can be taken as evidence that Equation 7-6 is valid at the larger times, and the slope of the line can be used to calculate the viscosity. Bird et al. also discuss the error introduced by the neglect of inertia and find that this error depends on the magnitude of the group, $[\rho h(dh/dt)/\eta]$.

Bird et al. ([1], p. 223) outline the derivation of the corresponding equation for a power law fluid. This is the "Scott equation."

$$F = \left(\frac{\pi k R^{n+3}}{n+3}\right)\left(\frac{2n+1}{2n}\right)\frac{(-dh/dt)^n}{h^{2n+1}} \tag{7-8}$$

Leider and Bird [291] have used the Scott equation to interpret the results of experiments on polymer solutions and have suggested a criterion to be used to decide when viscoelasticity will play a significant part in the fluid's response and thus render invalid the Scott equation. Lieder and Bird also consider the error introduced into the Scott equation by the neglect of inertia.

Brindley et al. [292] made a detailed theoretical analysis of squeeze film flow, and gave a method for the interpretation of the data when the form of the viscosity function is not assumed. They modified a Weissenberg Rheogoniometer for use as a squeeze film rheometer.

Shaw [293] used squeeze flow to characterize a series of ultra-high molecular weight polyethylenes. He used the Stefan equation to calculate an "apparent viscosity," but found that this varied with time. Thus, the value at a particular time (18 minutes) was selected as an empirical characterizing parameter.

Grimm [294] has considered the general question of the usefulness of squeeze flows for the study of polymeric liquids and has concluded that, at low speeds, they can be used to to determine viscosity. However, at higher speeds, there is no general method for treating the data, and the flow is not useful for the measurement of elastic properties.

We turn now to the case of constant volume flow in a parallel plate plastometer, shown in Figure 7-3. The theory for this flow is based on the same basic assumptions as above. Dienes and Klemm [295] have derived the following relationship between the force and the plate spacing for a Newtonian fluid, where U is the volume of the sample.

$$F = -\frac{3\eta U^2}{2\pi h^5}\left(\frac{dh}{dt}\right)$$

(7-9)

Integration yields

$$\frac{1}{h^4} = \frac{1}{h_0^4} + \frac{8\pi F}{3\eta U^2}t.$$

(7-10)

As in the case of squeeze flow, the initial stages of the flow cannot be expected to follow these equations, but Dienes and Klemm found that their plot of $(1/h^4)$ versus time was a straight line except at small times. Gent [296] has suggested that the assumptions regarding the velocity profile are valid only when $R > 10 \, h_0$, and he has given a correction to Equation 7-10 to be used when this condition is not satisfied.

Kataoka *et al.* [297] have used a parallel plate plastometer to study filled polymers. They used rheometrical equations derived by Oka and Ogawa [298] for use with power law liquids. They found this experimental technique to be useful only at shear rates below 0.2 s^{-1}.

Bartlett [299] and Bloechle [300] used a parallel plate plastometer to characterize glass-reinforced resins. The effect of the presence of glass fabric was explicitly taken into account in the analysis of the data.

Metzner and Denn [301] have performed a careful reappraisal of the assumptions made in the analysis of squeeze flows and found that one of these is questionable. This is the assumption that the wall pressure at the air-fluid interface is equal to the ambient pressure. They suggest that the correct boundary condition is that the normal stress at the free fluid surface is $-P_a$. The use of the correct condition makes the analysis of the flow considerably more difficult, but failure to do so can lead to significant errors in the case of non-Newtonian fluids.

ASTM standard Test Method No. D 926 is based on the use of a parallel plate plastometer. The plate spacing after a certain time is defined as the "plasticity number," and the recovery in h after the specimen is removed from the plastometer is called the "recovered height." A commerical instrument conforming to the ASTM specifications is described in Section 11.3.

7.6 CONVERGING FLOWS

In any flow involving the convergence of streamlines, the fluid will undergo stretching along these streamlines. Indeed, the steady simple extension described in Chapter 6 can be easily shown to be a converging flow by deriving the equations for its streamlines in a cylindrical coordinate system.

$$r^2 x = r_0^2 x_0 = c \qquad (7\text{-}11a)$$
$$\theta = \theta_0 \qquad (7\text{-}11b)$$

Equation 7-11 says that a fluid element that is located at the point (r_0, θ_0, x_0) at time, t_0, will be located at later times along a curve defined by these equations. A family of such streamlines for various values of r_0, for a single value of x_0, is sketched in Figure 7-4. The strain rate is uniform and is given by

$$\dot{\epsilon} = v_0/x_0 \tag{7-12}$$

where v_0 is the velocity in the x direction at $x = x_0$.

Another simply defined flow involving converging streamlines is three-dimensional sink flow in which the radial velocity is as follows in spherical coordinates.

$$v_R = \frac{-Q}{4\pi R^2} \tag{7-13}$$

The appearance of the streamlines in this flow is shown in Figure 7-5.

If we define $\dot{\epsilon}$ as the principal strain rate along a streamline, then

$$\dot{\epsilon} = \frac{dv_R}{dR} = \frac{Q}{2\pi R^3} \tag{7-14}$$

where Q is the "strength" of the sink; i.e., the total volumetric flow rate across any spherical surface. Clearly this flow is not equivalent to simple extension, as the strain rate is not uniform. Thus, the fluid elements do not undergo stretching at a constant rate as they move along, and this is not a motion with constant stretch history. Nonetheless, Marrucci and Murch [302] showed that, for steady sink flow, the only nonzero normal stress difference, evaluated at a particular value of R, can be expressed as a function of the local strain rate.

$$\tau_{RR} - \tau_{\theta\theta} = f(\dot{\epsilon}) \tag{7-15}$$

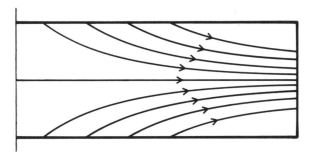

Figure 7-4. Shape of streamlines in uniaxial extension.

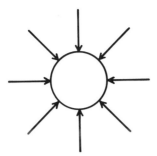

Figure 7-5. Sink flow streamlines.

However, the material function, $f(\dot{\varepsilon})$, can be related to the extensional viscosity only by assuming a particular constitutive equation. For example, in the case of a Newtonian fluid:

$$f(\dot{\varepsilon}) = 3\eta\dot{\varepsilon}. \tag{7-16}$$

Marrucci and Murch also analyze the case of two-dimensional sink flow.

7.6.1 Flow at the Entrance to a Capillary

As was explained in Chapter 4, the pressure drop associated with the flow of elastic liquids from a reservoir tube into a capillary was at one time thought to be associated primarily with recoverable shear strain. However, it is now recognized that this flow involves powerful stretching along streamlines, and it has been suggested that it be used as the basis for the design of a melt rheometer.

An important limitation to the rheometrical utility of converging flow at the entrance to a capillary, sometimes called "unconstrained convergence," is that it is not a controllable flow. This means that the flow pattern depends on the rheological properties of the fluid. For example, it has been observed for Newtonian fluids and certain polymers such as high density polyethylene and polypropylene that the convergence of the flow occurs over a short distance at the end of the reservoir, as shown in Figure 7-6a. On the other hand, for other polymers, such as low density polyethylene and polystyrene, the streamlines start to converge far upstream of the capillary entrance and take on a characteristic

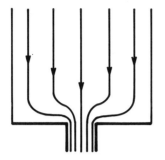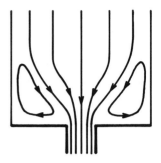

Figure 7-6. Flow at the entrance to a capillary: (a) without recirculation, (b) with recirculation.

"wine glass stem" shape as shown in Figure 7-6b. The melt outside the "stem" recirculates in a corner eddy.

A number of studies of the flow patterns of melts at capillary entrances have been reported, and White and Kondo [303] have reviewed much of this work. The main motivation for these studies was the desire to understand the cause of extrudate distortion or "melt fracture." Everage and Ballman [304], Hürlimann and Knappe [305] and Brizitsky *et al.* [306] found that there is a relationship between certain types of melt fracture and the extensional stresses generated in converging entrance flow. Cogswell [307] has summarized the observations of flow patterns and has also attempted to use converging entrance flows to measure rheological properties of molten polymers.

Cogswell [308, 221] has suggested a procedure for calculating a rheological property from measurements of entrance flows. Since neither the local normal stresses nor the local extension rate can be calculated directly from measurable quantities, it is necessary to make a number of assumptions about the flow. Cogswell assumes that the entrance pressure drop, ΔP_e, can be represented as a sum of two terms, one related to shear and the second to extension. He further assumes that the fluid follows streamlines that result in the minimum pressure drop. For a fluid having a power law viscosity function, he derives the following equations for an average extensional stress, σ_E, an average strain rate, $\dot{\epsilon}_E$, and an "apparent extensional viscosity," η_{EC},

$$\sigma_E = (3/8)\,(n+1)\,\Delta P_e \tag{7-17}$$

$$\dot{\epsilon}_E = \frac{4\eta(\dot{\gamma}_A)\dot{\gamma}_A}{3(n+1)\Delta P_e} \tag{7-18}$$

$$\eta_{EC} \equiv \sigma_E / \dot{\epsilon}_E = \frac{9(n + 1)^2 (\Delta P_e)^2}{32\eta(\dot{\gamma}_A)\dot{\gamma}_A^2} \qquad (7\text{-}19)$$

where $\dot{\gamma}_A$ is the apparent shear rate in the capillary whose radius is R.

$$\dot{\gamma}_A \equiv \frac{4Q}{\pi R^3}$$

Shroff *et al.* [309] used Cogswell's equation to interpret their entrance flow measurements involving several polymers, but they used the power law expression for the wall shear rate in place of the apparent (Newtonian) shear rate expression. They compared their results with the apparent extensional viscosity determined in a spinning experiment and found good agreement.

7.6.2 Lubricated Constrained Converging Flow

In his review paper [307], Cogswell suggested a method for generating uniform uniaxial extension by use of a specially shaped die as shown in Figure 7-7. A layer of low viscosity lubricant at the wall essentially

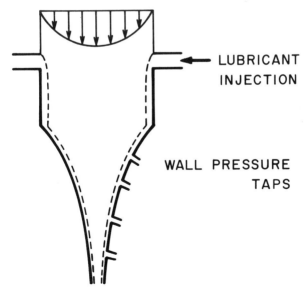

Figure 7-7. Converging flow rheometer proposed by Cogswell [307].

eliminates shear. Cogswell did not give the appropriate rheometrical equations, and these are not at all straightforward, especially the one relating stretching stress to wall pressures. It is useful to recognize that this is a radial compression rather than an axial extension. Thus, σ_{rr} has a larger magnitude than σ_{xx}, but since both are negative, this leads to a positive net stretching stress, $\sigma_{xx} - \sigma_{rr}$.

Shaw [310] used a die coated with silicone grease to study the role of extensional flow in melt fracture, but he did not try to calculate any rheological properties.

At about the same time that Cogswell proposed the use of lubricated convergence as a rheometrical flow, Everage and Ballman [311] reported the development of a rheometer based on the same principle. However, in their rheometer, the lubrication is provided by a flowing layer of "sheath fluid" which is not introduced until the entrance of the converging die. This introduces two important differences between this rheometer and the "ideal" one described above. First, the flow upstream of the die is not shear-free so that the flow is no longer of the extensional stress growth type from the point of view of a fluid element, and over some portion of the flow in the die, the melt is going through a transition from shear flow in the reservoir to extensional flow in the die. Secondly, the finite layer of the sheath fluid complicates the die design. Everage and Ballman did not measure wall pressures in the die, and reported a "strain-averaged" value of extensional viscosity. An equation for calculating the appropriate average stress was given without derivation. Data for one polyethylene were compared with values of average viscosity calculated from $\eta_T(t,\dot{\epsilon})$ data obtained by others using a uniform extensional flow, and good agreement was found.

Winter et al. [312] have proposed the use of a geometry that reduces the severity of entrance and exit conditions and which can also be used, in principle, to generate a wide variety of extensional flows. In "orthogonal stagnation flow," streams of the fluid under study flow in opposite directions, in two dies of the same design positioned on a common axis. The plane of symmetry between the two dies is a stagnation plane along which there are no drag forces. Depending on the shape of the dies and the direction of flow, all types of extensional flows can be generated if the melt can be prevented from adhering to the wall. Winter et al. [312] found it very difficult to provide suitable lubrication at the wall of the dies.

It should be noted here that in uniaxial extensional flow the stream surfaces are not shear free surfaces. This means that lubricated converging flow is not equivalent to uniaxial extension as it is defined in Equation 6-1. If the lubricant layer has negligible thickness, the streamlines at the wall can be made to conform to Equation 7-11, but the remaining streamlines will deviate from it.

7.6.3 Flow Toward a Collapsing Gas Bubble

Johnson and Middleman [313] have proposed a method for generating three-dimensional sink flow of a molten polymer and using this flow to determine an apparent extensional viscosity. The melt is contained in a closed, thermostatted metal box. A tube, projecting into the melt from the bottom of the box, is connected to a regulated gas supply system. A very sensitive pressure transducer monitors the pressure of the gas trapped above the melt and this indicates the volume of a gas bubble in the melt as a function of time. At the beginning of an experiment, a bubble of gas is introduced through the tube. Then the gas pressure is suddenly reduced to a level below the ambient pressure, and the bubble starts to collapse, generating the sink flow. The pressure of the gas in the bubble, P_B, and the volume of the bubble are measured as functions of time.

Turning now to the interpretation of the data in terms of a rheological material function, it is important to note that sink flow is not equivalent to uniaxial extension so that we cannot expect to establish a rigorous relationship between the function, $\eta_T^+(t,\dot{\epsilon})$, and quantities measured in this flow. For example, we note that the strain rate given by Equation 7-14 is not homogeneous, but varies with the radius.

Focusing attention on the liquid located at the liquid-bubble interface (i.e., at $R = R_B$), we note that the velocity is related to the bubble radius as follows.

$$v_R(R_B) = dR_B/dt \qquad (7\text{-}20)$$

From Equation 7-13, we have

$$Q = -\,4\pi R_B^2\,\frac{dR_B}{dt} \qquad (7\text{-}21)$$

and we can use Equation 7-14 to show that

$$\dot{\epsilon} = - 2(dR_B/dt)/R_B. \qquad (7\text{-}22)$$

We note that if the strain rate at R_B is to be maintained constant, then the bubble radius must have the following variation with time.

$$\ln R_B = -2\dot{\epsilon}t \qquad (7\text{-}23)$$

The determination of the normal stress difference $(\tau_{RR}-\tau_{\theta\theta})$ is more difficult. Pearson and Middleman [314] have considered this problem in detail. They show that for a Newtonian fluid,

$$\tau_{RR} - \tau_{\theta\theta} = - \tfrac{3}{2} \Phi \qquad (7\text{-}24)$$

where Φ is the difference between the bubble pressure and the hydrostatic pressure that would result in the liquid at the top of the bubble in the absence of motion. They further show that, for a Newtonian fluid,

$$\eta_T = 3\eta = \frac{\tau_{RR} - \tau_{\theta\theta}}{\dot{\epsilon}} = \frac{3\Phi R_B}{4(dR_B/dt)} . \qquad (7\text{-}25)$$

However, the relationship between the rheologically meaningful stress difference, $\tau_{RR}-\tau_{\theta\theta}$, and the measured pressure difference, Φ, depends on the constitutive equation of the liquid, so Equation 7-25 is not valid for non-Newtonian liquids. Pearson and Middleman suggest the use of an "apparent extensional viscosity" defined as follows.

$$\eta_{EB} \equiv \frac{3\Phi R_B}{4(dR_B/dt)} \qquad (7\text{-}26)$$

In their studies of molten polymers by use of the collapsing bubble technique, Johnson and Middleman [313] found that the maintenance of a constant pressure in the gas bubble tended to produce radius behavior following Equation 7-23 over most of the experiment. Thus, they interpreted their results in terms of plots of η_{EB} versus $\dot{\epsilon}$. Although there was considerable scatter in the points from different experiments, they seemed to follow the same type of behavior that has been observed in the case of $\eta_T(\dot{\epsilon})$ as measured in steady simple extension.

Münstedt and Middleman [315] compared values of η_{EB} with values of η_T obtained by use of the extensional rheometer described by Münstedt [247]. They found that, for mildly elastic melts (i.e., those with recoverable strains less than 1 and short relaxation times), the agreement was good. For a more elastic material, however, the agreement was unsatisfactory, with the two values differing by an order of magnitude at higher strain rates.

7.7 EXTRUDATE DRAWING (MELT SPINNING)

The drawing of an extruded filament of polymeric liquid is the principal step in the commercial manufacture of most synthetic fibers. For this reason, considerable work has been done on the simulation and analysis of this process. No effort will be made to review this work here, and the discussion will be limited to the use of extrudate drawing as a laboratory test for comparing polymer resins, whether or not they are fiber forming materials. This latter subject has been previously treated by Han ([6], Chapter 8), Hill and Cuculo [316] and Petrie ([21], p.93).

The basic features of the flow of interest are shown in Figure 7-8. The circular die or capillary is fed by an extruder or a piston moving in a reservoir, as in a capillary rheometer. The extruded filament is drawn down by means of a windup operating at constant speed. In its simplest form, the test involves the measurement of the tensile force at the windup device as a function of windup speed. This type of procedure is often called a "melt strength test," and it is described in more detail in Section 7.7.1. There has also been some effort made to modify this simple test to make possible the measurement of a material function, and this work is described in Section 7.7.2.

7.7.1 The Melt Strength Test

An early report of the use of extrudate drawdown to characterize molten polymers is that of Busse [317]. As the windup speed was increased, he found that the tension in the filament increased quite rapidly at first and then leveled off, although there were large fluctuations in the tension, especially at the highest speeds used. Ultimately, the filament broke. The maximum, time-averaged value of the tension was called the "melt strength," and Busse found that this was related to the low shear rate viscosity, η_0, for a given class of polymers. The filament was allowed

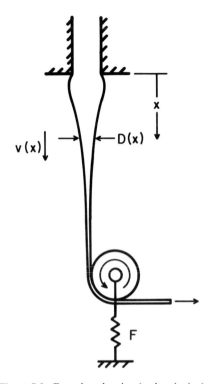

Figure 7-8. Extrudate drawing (melt spinning).

to cool in air prior to windup so that the drawdown did not occur isothermally.

Cogswell [318] submerged the windup in a quench bath, and used an empirical equation to relate experimentally measured quantities to the extensional viscosity, $\eta_T(\dot{\epsilon})$.

Meissner [231, 319] modified the Busse apparatus by using a rotary clamp as the drawdown mechanism.* This device, described in Section 6.1.3, allows tension to be applied directly to the melt without the necessity of cooling the sample below its solidification temperature. Meissner [231] found that the use of a melt tensile tester provided a basis for distinguishing between several film blowing resins that gave similar results in shear tests but differed markedly in their processing behavior.

Wissbrun [320] investigated the possible relationship between melt strength and well-defined rheological material functions. He concluded

---

*A commercial test apparatus based on this principle, the "Rheotens," is described in Chapter 11.

that melt strength depends on a number of more fundamental properties and on the cooling rate, and that the melt strength test cannot be used to determine the extensional viscosity of a melt.

Swerdlow et al. [321] designed a simple quality control test for polyethylene film resins that is a modified version of the melt strength test. Weights of successively increasing mass are attached to the extrudate from an extrusion plastometer, and the highest load that the extrudate can withstand before breaking is used to calculate a rupture stress. Swerdlow et al. report a correlation between rupture stress and drawdown failure in the film blowing process.

7.7.2 Isothermal Melt Drawing

In an effort to obtain more meaningful rheological data, the melt strength test has been modified by extruding into a chamber maintained at the extrusion temperature and measuring the filament diameter or local velocity as a function of distance from the die. Fehn [322] used this technique and interpreted his results in terms of a time-dependent tensile modulus. The local velocity was measured by means of tracer particles in the melt, and the tensile stress in the melt was calculated from the measured tensile force by including the effects of acceleration and gravity. The strain was computed as follows,

$$\epsilon(t) = [v(t) - v_0]/v_0 \qquad (7\text{-}27)$$

where $v(t)$ is the local velocity of the filament at a point where the melt has been traveling for a time t since exiting the die. The quantity, v_0, is the velocity which would occur at the die exit if the extrudate relaxed instantaneously upon leaving the die; it must be determined indirectly.

Acierno et al. [323], Han and Lamonte [324] and Deprez and Bontinck [325] measured the diameter, $D(x)$, as a function of distance from the die in isothermal experiments and calculated the velocity, $v(x)$, by use of the continuity equation:

$$Q = v(x) [D(x)]^2 4\pi \qquad (7\text{-}28)$$

where Q is the volumetric extrusion rate. Then the local strain rate was calculated as follows.

$$\dot{\epsilon}(x) = dv/dx \qquad (7\text{-}29)$$

If inertia, gravitational force, air drag and surface tension are neglected, the tensile stress is directly related to the measured filament tension, F.

$$\sigma_{xx} = F/(\pi d^2/4) \qquad (7\text{-}30)$$

An "apparent extensional viscosity" can then be calculated as follows.

$$\eta_{ES} \equiv \sigma_{xx}/\dot{\epsilon}(x) \qquad (7\text{-}31)$$

However, since the flow does not involve a constant stretch history, this apparent extensional viscosity is not a material function, but depends on the details of the experiment such as the flow rate, Q.

Bayer et al. [326] found that when they plotted the apparent extensional viscosity as a function of t rather than ϵ, where t is the time a fluid element has been moving since leaving the die, a universal curve was obtained by all values of Q. This universal curve of $\eta_{ES}(t)$ was a straight line except at very short times. This led Bayer [327] to propose a procedure for interpreting the data in terms of three material constants.

Spearot and Metzner [328] used an interesting technique to determine the local stress in a molten filament. They extruded horizontally into a chamber maintained at the extrusion temperature. After leaving the chamber, the filament passed over a support to the windup where tension was applied. The filament sagged between the die and the support, and the local tensile stress was related to the local slope of the filament. They found that the diameter tended to vary with distance along the filament as

$$1/D^2 = Ax + B. \qquad (7\text{-}32)$$

This implies that the strain rate is constant along the filament.

Chen et al. [329] made an experimental and theoretical study of the meaning of the apparent extensional viscosity function. They concluded that the inevitable complexity of the strain history, both in the die and just after extrusion, affects the behavior of the melt in the filament so that a universal, purely extensional deformation history cannot be generated in practice.

7.3.3 The Usefulness of Spinning Experiments

The melt strength test is attractive to the plastics engineer because it subjects the melt to a deformation that is primarily a uniaxial elongation, without requiring the use of the complex apparatus and techniques described in Chapter 6. However, even if the test is modified by providing for isothermal drawing, the flow has neither a constant nor a controlled strain history, and the data cannot be interpreted in terms of well-defined material functions. Shearing in the die and extrudate swell influence the behavior of the melt after it enters the zone of uniaxial extension and this precludes the possibility of a universal strain history.

Thus, the utility of tests based on extrudate drawing is limited to the comparison and screening of resins from a single family of polymers, and the determination of well-defined material functions is not possible.

7.8 FILM BLOWING

Han and Park [330] and Iwakura, Yoshinari and Fujimura [331] have suggested the use of a small-scale, isothermal version of the commercial film blowing process to measure some kind of extensional viscosity of molten polymers. However, this is an asymmetrical, nonhomogeneous stretching flow in which measurement of local stresses is not possible. Although local strain rates can be measured, as demonstrated by Farber and Dealy [332], the strain history is neither constant nor controllable, so that interpretation of measurements in terms of a well-defined material function is not possible.

Han and Park [330] calculated an "apparent extensional viscosity" and plotted it as a function of local strain rate. However, this representation ignores the role of the strain history (i.e., the elasticity of the melt) and it is difficult to see how any absolute significance can be attributed to it.

Chapter 8
How To Select a Melt Rheometer

In selecting an instrument for a particular application, it is useful to begin by making sure that certain basic questions have been answered. For example, one must know why the data are needed and what kind of personnel will be operating the equipment. Basic questions of this sort are discussed in Section 8.1. Once these questions have been answered, it is necessary to make a series of specific decisions that will lead to the identification of the instrument best able to meet one's needs. Section 8.2 contains advice on the making of such decisions.

8.1 BASIC CRITERIA

Why are rheological data needed?

This is the first and most important question that one must ask in selecting a rheometer. If the data are to be used for quantitative resin characterization, a precise measurement of a well-defined material function will probably be required. For example, the dynamic viscosity and storage modulus are often used as indicators of differences in molecular structure. On the other hand, if the measurements are to be made for routine quality control, it will be desirable that the testing procedure be simple and rapid. Rheological properties are often included in contract specifications for the sale of resins. When this is the case, the acceptance of a shipment may depend on the outcome of the measurement, and a high degree of automation may be justifiable in spite of the much higher cost.

On-line measurement of samples from a flowing stream is sometimes used for quality control or even for automatic process control. If this is the goal, the choice of instrument will be quite limited, as it would appear that only Seiscor and Göttfert manufacture this type of equipment.

The comparison and screening of resins by a resin manufacturer or large scale processor may require more elaborate measurements than those used for quality control, and if one wishes to establish values for the material constants of a constitutive equation to be used for detailed process modeling, then a fairly complete characterization will be required.

In what process is the resin to be used?
The answer to this question may govern the specific test performed or property measured. This point will be discussed more fully in Section 8.2.2.

Where is the rheometer to be located?
Some instruments are fairly robust and can be used in the plant for routine testing, and some of these are portable and can be easily moved from one location to another. On the other end of the spectrum are sophisticated instruments containing components highly sensitive to vibration, heat and moisture. Such devices often require specialized services and environments. For example, high pressure, filtered and dried air might be required to operate an air bearing. It is not sensible to pay $100,000 for a scientific instrument and then not provide the environment necessary for it to yield reliable data.

What funds are available?
Melt rheometers range in price from $2,000 to $250,000. Installation and maintenance costs also vary widely. It is to be expected that the more sophisticated devices will require repairs and adjustments at least once or twice a year, and budgetary provision should be made for this.

Who will operate the rheometer?
Some skill is required to operate any test equipment if reliable results are to be obtained. But the more elaborate rheometers will require a high degree of skill and some basic understanding of rheology on the part of the operator. If such personnel are unavailable, it is probably a mistake to invest in a complex and costly rheometer.

Are facilities available for fabrication or modification of a rheometer?
If no such facilities are available, then an instrument must be purchased that is closely suited to the exact needs of the user.

How urgent is the need?
Delivery times vary depending on many factors. A standard model of
a simple device may be available from stock. In other cases, delays of
six to ten months may be involved.

8.2 THE SELECTION PROCESS

There are several decisions that must be made once the above questions
have been answered.

Standard test or property measurement. A standard test is based on
an arbitrarily defined procedure, the results of which cannot be interpreted in terms of fundamental polymer properties. Examples are the
melt index and Rossi-Peakes tests. The latter of these is an example of
a test which simulates in some sense an industrial process—in this case,
molding. Unfortunately, simple scale-up principles have not been established for these processes so that there is no quantitative way to relate
the test result to processability.

Although they do not yield values of physical properties, standard
tests are usually easy to perform using relatively inexpensive equipment
and are often adequate for routine quality control work or for the comparison or screening of resins from a single family of polymers (i.e.,
polymers produced by a given type of reaction), but with minor variations in reactor conditions.

A case in point can be cited from the author's experience. A profile
extrusion specialist expanded his business rapidly to supply high density
polyethylene pipe for a new market. The pipe was being delivered and
installed as fast as it could be produced. A melt indexer was available,
but all available personnel were assigned to production and no testing
was done. Routine pipe inspections revealed some variation in quality,
but realization that serious flaws were present came only when large
quantities of pipe failed in field tests and in service. These failures led
ultimately to the bankruptcy of the processor. It was later ascertained
that this process was one in which the optimum operating conditions
were very sensitive to resin formulation and that the resin supplier had
periodically changed the formulation. Furthermore, a simple melt index
test would probably have been sufficient to identify batches of resin
likely to require the modification of the operating conditions.

Measurement of a well-defined rheological property, on the other
hand, requires the use of a "controllable" flow in which both stress and

strain are known. Instruments for property measurement are generally more complex than those used for standard tests but yield information of more fundamental significance, which can be used to compare and classify resins of different types. Single point standard tests are generally inadequate for this type of work. For example, as was illustrated in Figure 4-13, it is possible for several resins to have similar values of melt index and have at the same time a wide diversity in their fundamental rheological properties; e.g., the shape of the viscosity-shear rate curve.

Which test to perform, which property to measure. Table 8-1 lists all the test methods described in this book, while Table 8-2 summarizes all the measurable material functions that have been defined. These tables also indicate the sections in which one can find a description of the test or property and descriptions of relevant commercial instruments.

Certain tests have come to be associated with certain materials. For example, specification sheets for thermoplastic resins often include the melt index, as measured in an extrusion plastomer. In the case of PVC, compounding and fusion are usually characterized by use of a torque rheometer. When one is dealing with materials like these, it is usually useful to be able to carry out the traditional test. However, just because a particular test method is commonly used is no reason to assume that all aspects of processing behavior can be correlated with the results of this single test.

Table 8-1. Test Methods for Molten Polymers

TEST	TYPICAL APPLICATION	ASTM STANDARD METHOD	SECTION IN WHICH FLOW IS DESCRIBED	SECTION IN WHICH COMMERCIAL INSTRUMENTS ARE DESCRIBED
Extrusion plastometer	Extrusion	D1238	4.4.1	9.2
Rossi Peakes flow tester	Molding	D569	4.4.1	9.3
Mooney Viscometer (Shearing disk viscometer)	Elastomer processing	D1646	5.4	10.8.3
Parallel plate plastometer	Rubber processing	D926	7.5	11.3
Squeeze film flow			7.5	11.3
Torque rheometer	Mixing, extrusion	D2538	7.3	11.4
Converging flows	Flow into dies		7.6	
Extrudate swell	Profile extrusion		7.4	9.4.3, 9.5.1
Melt spinning	Fiber spinning		7.7	11.1

Table 8-2. Measurable Material Functions

NAME OF MATERIAL FUNCTION	SYMBOL	SECTION WHERE DEFINED
Time-independent material functions		
Viscometric functions	VMF	2.3.1
Viscosity	$\eta(\gamma)$	
First normal stress difference	$N_1(\dot{\gamma})$	
Second normal stress difference	$N_2(\dot{\gamma})$	
Extensional viscosities		
Uniaxial	$\eta_T(\epsilon)$	2.3.3
Planar	$\eta_p(\dot{\epsilon})$	2.3.5
Biaxial	$\eta_b(\dot{\epsilon}_b)$	2.3.4
Linear viscoelasticity functions (shear)	LVF	
Stress growth function	$\eta^+(t)$	2.4.2
Relaxation after steady shear	$\eta^-(t)$	2.4.2
Creep compliance	$J(t)$	2.4.3
Dynamic viscosity	$\eta'(\omega)$	2.4.4
Storage modulus	$G'(\omega)$	2.4.4
Stress relaxation modulus	$G(t)$	2.4.1
Nonlinear viscoelasticity functions	NLVF	
Shear		
Stress growth function	$\eta^+(t,\dot{\gamma}_0)$	2.4.2
Relaxation after steady shear	$\eta^-(t,\dot{\gamma}_0)$	2.4.2
Creep compliance	$J(t,\sigma_0)$	2.4.3
Superposed steady and oscillatory	$\eta'(\omega,\dot{\gamma}_m)$	2.4.5
shear	$G'(\omega,\dot{\gamma}_m)$	
Extension		2.4.7
Stress growth function	$\eta_T^+(t,\dot{\epsilon}_0)$	
Relaxation after steady extension	$\eta_T^-(t,\dot{\epsilon}_0)$	

It is essential to keep in mind that no single measurement gives a complete rheological characterization of a material as complex as a molten polymer. Each property gives a different perspective on the flow behavior, and the challenge for the rheologist is to select the property that is most appropriate for a particular application. Direct quantitative links between test results and processing behavior can be established by means of mathematical modeling only in the case of a few relatively simple processes such as extrusion.* In general, such links can only be established on the basis of a systematic series of carefully designed experiments.

---

*In this case, it is the $\eta(\dot{\gamma})$ function that governs the flow behavior of the melt in the extruder (as shown, for example, by Middleman [10]).

Because the number of properties that could be measured was, until fairly recently, quite limited, such experimental programs often failed to yield useful results. For example, Miller [333] has reported on the results of a major effort in an industrial research and development laboratory to establish a relationship between certain aspects of parison behavior in the blow molding process and several rheological properties. The properties considered were viscosity as a function of shear rate $\eta(\dot{\gamma})$, and the dynamic properties, $\eta'(\omega)$ and $G'(\omega)$. It was found that the parison swell could not be correlated with any of these properties.

Dealy and Garcia-Rejon [334] were also unable to establish any simple relationship between parison swell and readily measured properties such as viscosity and capillary extrudate swell. However, they found that parison sag was closely related to the extensional stress growth function, $\eta_T^+(t,\dot{\epsilon})$.

Another situation in which an extensional flow test was found useful has been described by Meissner [231]. He found that differences in the processing behavior of several film blowing resins could not be associated with differences in the viscometric and linear viscoelastic properties but that the melt strength test was successful in differentiating between these resins.

In the absence of a specific precedent or an established correlation, it would seem logical to select a test involving the type of deformation thought to be of central importance in the process of interest. Once the type of deformation has been selected, it is important to carry out the test using temperatures and strain rates that are as near as possible to those that occur in practice. Münstedt [335] has suggested the following as the shear rate ranges relevant for each of the processes listed.

Process	*Shear Rate, s^{-1}*
Mechanical shrinkage	10^{-3} to 10^{-1}
Pressure molding	10^{-1} to 10
Extrusion	10 to 10^3
Injection molding	10^3 and higher

While no single rheometer can be used over this entire range, every shear rate indicated is accessible in some commercial instrument. For example, a rotational machine with cone plate geometry can cover the low end, up to about 10 s^{-1}, while a capillary rheometer can be used to measure viscosity at the higher shear rates.

Finally, strain history should be taken into account, particularly if the process involves rapid changes in deformation patterns or strain rate. For example, we might note that the melt spinning process is primarily a uniaxial extensional flow and conclude that the extensional viscosity, $\eta_T(\dot\epsilon)$, would be related to spinning behavior. In assessing this possibility, we first note that none of the techniques described in Chapter 6 for determining $\eta_T(\dot\epsilon)$ are useful at the high extension rates typical of melt spinning. In addition, the spinning process involves a complex strain history in which, starting in the die, there is shear, recoil (swell), and finally a uniaxial stretching at a variable rate. Obviously, no single property will provide a basis for a description of all aspects of such a process.

Method of generating the deformation. If a property is to be measured, there may be an option as to the specific flow "geometry" or type of apparatus used. For example, viscosity can be determined by the use of a wide variety of capillary and rotational rheometers. All the test geometries described in this book are listed in Table 8-3, together with sections in which they are described.

Extent to which automation is required, The most recent trend in commercial rheometers is toward increased automation. This adds substantially to the cost and complexity of the apparatus, but reduces the time and effort required to carry out a test. Dedicated minicomputers and built-in microprocessors are often available to provide a direct, digital readout or printed record of the desired property, eliminating the manual tabulation and analysis of data. These aids are likely to be a good investment when a large number of tests must be performed.

Unfortunately, there has been little progress in the automation of the most time-consuming aspects of melt rheometry, sample preparation and cleaning.

Precision required. For the reasons outlined in Chapter 2, rheological data for melts are not as reproducible as those for low molecular weight liquids. Variations between runs of 5% cannot be considered as significant, and for many purposes 10% accuracy may be acceptable. Variations from one batch to another of resin produced under the same nominal conditions can be somewhat greater than this and would thus be detectable even at this level of precision.

Buy or build. Those having access to sophisticated machine fabrication facilities may wish to design and build their own rheometers, and they may find some useful ideas in Chapters 4 through 7. Most industrial rheologists, however, do not have the time to develop new rheo-

Table 8-3. Flow Geometries for Property Measurement

	Section in Which Described	Chapter in Which Commercial Instruments are Described	Properties That Can be Measured
Sliding plate[a]	4.1		$\eta\ (\dot{\gamma})$, LVF
Sliding cylinder[a,b]	4.2		$\eta\ (\dot{\gamma})$, LVF
Falling cylinder[b]	4.2		$\eta\ (\dot{\gamma})$
Capillary[c]	4.3,4.4,4.6	9	$\eta\ (\dot{\gamma})$, $N_1\ (\dot{\gamma})$, $N_2\ (\dot{\gamma})$
Slit[c,d]	4.5,4.6	9	$\eta\ (\dot{\gamma})$, $N_1\ (\dot{\gamma})$
Annulus (axial flow)[e]	4.7		$N_2\ (\dot{\gamma})$
Annulus (circumferential flow)[e]	4.7		$N_1\ (\dot{\gamma})$
Concentric cylinder[f]	5.1	10	$\eta\ (\dot{\gamma})$, LVF, NLVF
Cone and plate[f,g]	5.2	10	VMF, LVF, NLVF
Bicone[f]	5.2.4		$\eta\ (\dot{\gamma})$,LVF, NLVF
Parallel disks (torsional flow)[f,g]	5.3	10	$\eta\ (\dot{\gamma})$, $N_1 + N_2$
Eccentric rotating disks[f,g]	5.5	10	$\eta'(\omega)$, $G'(\omega)$
Uniaxial extension	6.1	11	$\eta_T(\dot{\epsilon}),\eta_T^+\ (t,\dot{\epsilon}),\eta_T^-\ (t,\epsilon_0)$
Planar extension	6.3		$\eta_p(\dot{\epsilon}),\eta_p^+\ (t,\dot{\epsilon})$
Biaxial extension	6.2		$\eta_b(\dot{\epsilon}_b),\eta_b^+(t,\dot{\epsilon}_b)$

[a]Limited to use at low maximum strain.
[b]Limited to use at low strain rate.
[c]Use of exit pressure to determine normal stress difference depends on assumptions which have not as yet been decisively verified.
[d]Use of pressure hole error to determine normal stress difference depends on assumptions that have not as yet been decisively verified.
[e]Form of equation for normal stress function must be assumed.
[f]Maximum shear rate limited by edge effects.
[g]Maximum shear rate limited by tendency of fluid to flow out of gap.

metrical methods and will wish to purchase a commercially manufactured instrument.

Which instrument to buy. Chapters 9, 10 and 11 contain information useful in the selection of a commercial rheometer. It was not possible to buy and test each instrument described as is done by some consumer testing laboratories. Therefore, the data given are limited to those supplied by the manufacturers. Price codes refer to the ranges listed in Appendix C. Since detailed specifications and prices can change at any time, the manufacturer or his agent should be consulted for the latest information.

It is always wise, before issuing a purchase order, to follow the advice of the now defunct Packard Motor Car Co.: "Ask the man who owns one."* Manufacturers often have available lists of previous purchasers of their equipment.

---

*Of course, today this would have to be amended: "Ask the *person* who owns one."

Chapter 9
Commercial Testers and Rheometers Based on Capillary and Slit Flow*

9.1 INTRODUCTION

This presentation of information on capillary flow instruments is organized on the basis of the method of generating the driving pressure. Driving mechanisms used in commercial devices include the following.

A. Weight-driven piston
B. Gas pressure
   1. Direct to barrel
   2. Piston driven by pneumatic cylinder
C. Electric motor–Piston
   1. Electromechanical drive
      (motor linked mechanically to piston)
   2. Hydraulic drive
      (motor operates hydraulic pump; oil pressure drives piston)
D. Melt pump
   1. Positive-displacement, melt-fed pump
   2. Extruder

With the exception of Category D, this list is organized in order of increasing cost, complexity and versatility. The weight-driven devices are basically testers and are not well suited to accurate property measurement. The instruments in Categories B and C, as well as some of those in D, are true melt rheometers allowing the determination of the viscosity function. The Seiscor/Han slit rheometer, described in Section

---

*See Appendix C for price range key used throughout this chapter.

9.8, provides also an estimate of the first normal stress difference based on the exit pressure concept.

Capillary rheometers fed by extruders are usually offered in the form of accessories for use with torque rheometers, and for this reason they are described in Chapter 11. An exception is the Göttfert Minex Rheometer, which is described in Section 9.2.

9.1.1 Estimation of Shear Rate and Shear Stress Ranges for Instruments

For a complete presentation of the rheometrical equations for capillary and slit flow, as well as sources of error, the reader is referred to Chapter 4. It is useful, however, to review a few of the basic equations here to show how to interpret manufacturers' specifications in terms of rheological variables. The design specifications for capillary rheometers are expressed in terms of such quantities as die dimensions, driving pressure, load cell capacity and piston speed range. The variables of rheological significance, on the other hand, are shear rate, shear stress and viscosity.

The shear stress at the wall is related to the driving pressure as shown by Equation 4-28. The end correction depends on the rheological behavior of the material as well as the detailed geometry of the entrance to the die. However, if the L/D ratio is sufficiently large, say 15 or greater, the end correction will be less than 10% of this, so an approximate value of the wall shear rate is ($RP_d/2L$).

If the driving pressure is determined by means of a pressure transducer in the barrel, then the available range of driving pressures will be governed by the range of the transducer, presuming, of course, that the driving mechanism can produce this range of pressures and that the rheometer is strong enough to support the maximum pressure involved. Equation 4-26 gives the relationship between the plunger force and the driving pressure.

The shear rate at the wall is the second basic rheological quantity, and it is related to the volumetric flow rate as shown in Equation 4-19. However, this equation involves a derivative of flow rate with respect to pressure that cannot be determined *a priori,* without assuming a specific form for the viscosity function, such as the power law. Thus, given only the rheometer design parameters, one cannot calculate the accessible range of shear rates. It is customary in this case to use the apparent wall shear rate as an estimate of the true value, although, as pointed out in Section 4.3.1., the discrepancy can be rather large. The apparent wall shear rate is the wall shear rate of a Newtonian fluid at the same flow rate. In terms of the piston (plunger) speed, V_P, and the radii of the barrel, R_b, and capillary, R, the apparent shear rate, $\dot{\gamma}_A$, is given by Equation 9-1.

$$\dot{\gamma}_A = \frac{4Q}{\pi R^3} = \frac{4V_P R_b^2}{R^3} \qquad (9\text{-}1)$$

The shear rate ranges given in manufacturers' specifications are based on this equation and are thus estimates that can be in error by as much as 50%, depending on the extent to which the melt studied is non-Newtonian.

Rheometer specifications often include the range of viscosities that can be measured. This range is calculated by use of Equation 9-2, which neglects the entrance correction and makes use of the apparent shear rate.

$$\eta = \frac{\pi R^4 P_d}{4QL} \qquad (9\text{-}2)$$

Thus, it represents a rough approximation of the viscosity range measurable for any actual melt. Furthermore, it should be noted that the highest measurable viscosity corresponds to a combination of the largest P_d and R and the smallest Q value. In other words, this maximum viscosity is only measurable at the lowest shear rate. Likewise, the lowest viscosity measurable corresponds to the highest shear rate.

9.2 EXTRUSION PLASTOMETERS (MELT INDEXERS)

The extrusion plastometer, or "melt indexer," as it is often called, is the most popular melt tester now in use. Its design, based on the work of John Tordella and others at the E. I. DuPont Experiment Station, grew out of the need for a simple, reliable method for characterizing molten thermoplastics. It has largely displaced older testers because of increased versatility and improved control of variables. For reasons given in Section 4.4.1, it is not well suited to the precise determination of viscosity, nor to the comparison of resins of different families. But for routine quality control, it is a popular choice because it is inexpensive to buy, robust and simple to use, and requires only electricity in the way of service utilities.

Evidence of the worldwide popularity of the extrusion plastometer is provided by the fact that it is the basis of at least eight national and international standard test methods as shown below.

ASTM D1238-73	U.S.A.
BS 2782-105C	United Kingdom
DIN 53735	West Germany
AFNOR T51-016	France
NFT 51-006, 51-016	Netherlands
UNI 5640-74	Italy
JIS K7210	Japan
ISO R1133, R292	International

The test instruments described in most of these various standards are nearly identical so that the "standard" melt plastometer is an instrument of international importance.

As explained in Section 4.4.1, the standard test involves the determination of the mass of polymer extruded in 10 minutes. The standard die has a diameter of 2.095 millimeters and a length of 8 millimeters. In the ASTM test method (D 1238), and in some of the others listed above, two methods are described for measuring the amount of polymer extruded. In "Procedure A," the extrudate is allowed to accumulate at the exit of the die and simply cut off and weighed at the end of 10 minutes. For materials with large flow rates (i.e., those with low viscosities) or when automation is desired, "Procedure B" provides for the use of a switch and timer that automatically determine the time required for the piston to fall through a specified distance. The standard "flow rate" is then calculated as follows.

$$\text{Flow rate (grams/10 minutes or dg/minute)} = KL\rho/t \quad (9\text{-}3)$$

where
K = a constant, depending on the effective barrel diameter*
L = distance through which piston falls in time t
ρ = density of melt

---

*For the standard instrument dimensions and if L is expressed in centimeters and t in seconds, K is equal to 427.

Although there is only one standard geometry, no fewer than 20 combinations of temperature and load are specified to accommodate a wide variety of materials. Condition "E," often used for polyethylene, involves a temperature of 190°C and a load of 2.160 kilograms, from which the nominal driving pressure can be calculated to be 298.2 kPa. Of course, the actual driving pressure will differ somewhat from this value because of piston friction and the weight of the polymer in the barrel.

Melt plastometers designed in accordance with one or more of the various standard test methods are manufactured by at least 12 companies in 5 countries.* Basic information on these instruments is tabulated in Table 9-1. Prices start in the "C" range for a basic unit with relay temperature control and suitable for operation up to 200°C. Higher priced units generally have more sophisticated temperature control and higher maximum operating temperatures. Models with proportional temperature control and capable of operation up to 300°C are priced in the "D" range. A model with reset-integral-derivative temperature control and a digital temperature indicator is available in the "E" range.

Some manufacturers offer certain added features. A piston travel sensor with switch and timer, to permit the use of Procedure B of the ASTM standard method, is available on almost all models at a price in the "B/C" range. A plunger guide and a weight lift device, either manual or pneumatic, are very useful when using the standard test conditions involving heavy loads. (The largest standard load is 21.6 kilograms.) Weight lift mechanisms begin in the "A" price range. Other options offered include built-in manual or automatic extrudate cutting devices, piston travel versus time recorders, and microprocessors for automatic calculation and display of the melt flow.

Several of the models listed in Table 9-1 involve some degree of automation. Option "H," for example, which is available on several models, involves a microprocessor which receives a signal from a piston travel timer and calculates the melt flow by use of Equation 9-3. The resulting value appears on a printed paper tape.

In the Kayeness Model D-9052, it is possible to preprogram a sequence of five successive tests, all at one loading, each involving the

---

*The first commercial plastometers were made by the Slocumb Co.

Table 9-1. Commercial Extrusion Plastomers ("Melt Indexers")

MANUFACTURER (U.S. MARKETING AGENT)	MODEL	TEMPERATURE CONTROL SYSTEM	MAXIMUM TEMPERATURE	STANDARD FEATURES	OPTIONS	NOTES
CEAST	6540/000	Electronic	400°C	E	A,B,D,H,F,N,O	
(Intermetco)						
Custom Scientific	CS-127	Electronic	400°C	G	B,D	
	CS-127 MF	Electronic	400°C	G	B,D	Q
	CS-127 A	Electronic	400°C	G,J	B,D	
Davenport	"Utility"	Hg-Relay	275°C			
(TMI)	Model 3	Pt-Relay	275°C		D	
Karl Frank	73592	Electronic	400°C		I,G	
	73593 SW	Electronic	400°C		B,D,F,G,I,K	
Göttfert	Melt Indexer	Electronic	400°C		B,C,D,F,G	
(Automatik U.S.A.)					H,I,J,K,N,O	
	Side Stream	Electronic	400°C		J,K	H,L,M
Kayeness	D-7050	Electronic	300°C	A	C	
	D-7051	Electronic	300°C	A,D,	C	
	D-0053	Electronic	300°C	D,H,J,O,S	C	
	D-9052	Electronic	300°C	D,H,J,O,S		U,V
Ray-Ran	Mk 1 Utility	Electronic	300°C	A,N		
(TMI)	Mk 2 Digital	Electronic	300°C	A,N,O		
Monsanto	3504		300°C		P	
Seiscor	AK-2000	Electronic	300°C			M
Slocumb	Models A,B,C	Hg-Relay	400°C		D	R
	Models D,E,F	Electronic	400°C		D	R
Tinius Olsen		Electronic	400°C	O	A,B,D,S	
Toyoseiki	522		300°C		D	
	592	Electronic	230°C	C,D,F,H		L
Wallace	P6	Hg-Relay				
(TMI)						
Zwick	4104	Electronic			B,D,F,G,H, I,N,T	

[A]Built-in timing device (electric stopwatch).
[B]Manual weight lift device.
[C]Pneumatic weight lift device.
[D]Piston travel sensor switch and timer.
[E]Manual extrudate cutting mechanism.
[F]Automatic extrudate cutting mechanism.
[G]Weight guide.
[H]Minicomputer and printer for calculation and display of melt flow.
[I]Flow recorder (piston height versus time).
[J]Flow timer and automatic preheat cycle timer with electronic readout.
[K]Automatic cycling through four loads.
[L]Completely automatic operation.
[M]In-line melt index simulator—described in Section 9-7.
[N]Platinum resistance thermometer.
[O]Digital temperature display.
[P]180°F to 1000°F temperature range.
[Q]"Micro-flow" miniature plastometer for small samples.
[R]Maximum temperatures of 200°C, 270°C and 400°C for the three models.
[S]Electronic display of elapsed time.
[T]Extrudate diameter detector.
[U]Pneumatic cylinder replaces dead weight loads.
[V]Automatic cycling through four tests with programmable delays between tests.

travel of the piston through a distance of ¼ inch with preselected delay times between tests. Thus, a single loading of the barrel yields five successive values of the time required for the piston to fall ¼ inch, so that the time dependency of the flow can be observed. These values are accumulated in a microprocessor, which then calculates the melt flow by use of Equation 9-3. Also calculated are the apparent shear rate, the average melt flow for the five tests and the standard deviation. This is actually not a weight-loaded device, and the load is applied by means of a pneumatic cylinder. The force on the piston is monitored by means of a load cell. The Model D-9052 is priced between the "F" and "G" ranges.

The Toyoseiki Model No. 592 is unique in the extent to which its operation has been automated. The operator simply places 4 grams of resin into a traveling specimen bucket. Melting, degassing, extrusion and cleaning proceed automatically, and the melt flow rate is calculated according to Equation 9-3 and appears on a printed tape.

Another instrument designed for automatic determination of melt flow is the Seiscor Continuous Melt Rheometer. This device is intended for continuous on-line service in resin production and melt processing. It is not actually a weight-driven tester, but simulates in some sense the extrusion plastometer and yields data said to be equivalent to standard melt flow rates. The actual driving force for the flow is a melt-fed metering pump and, for this reason, the Seiscor instruments are described in Section 9.7.1 rather than in the present section.

In the Minex Rheometer made by Göttfert, shown in Figure 9-1, the weight-driven piston is replaced by a small extruder, which both melts the resin and pumps it through a capillary. A transducer senses the pressure at the entrance to the die. The standard circular die can be replaced by a slit die with two wall pressure transducers. The manufacturer's specifications imply that the flow rate can be determined from the screw speed, but as an extruder is not a positive displacement pump, it may be necessary to verify the flow rate by collecting extrudate.

Some of the features offered on the various commercial extrusion plastometers can be seen in Figure 9-2 to 9-5. Figure 9-2 shows the basic instrument manufactured by Custom Scientific. The two knobs and voltmeter mounted on the front panel are related to the temperature control system. A glass thermometer, bent at a right angle, can be seen at the top of the cylindrical housing for the insulated barrel and

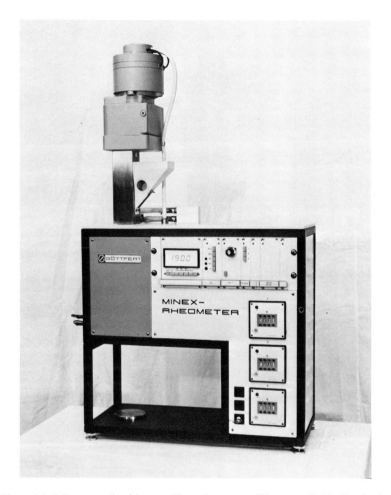

Figure 9-1. Minex extruder-driven capillary plastometer. (Photo supplied by Göttfert.)

die. Figure 9-3 shows the Kayeness Model D-7051, which has a piston displacement sensor and a built-in timer.

Figure 9-4 shows a CEAST instrument equipped with a digital temperature indicator, an electronic timer, an extrudate cutting device, a mechanical weight-lifting mechanism, a piston travel sensor and a calculator. The calculator is used in conjunction with Procedure B and operates on the basis of Equation 9-3. The product ($KL\rho$) is entered manually and stored. After each run, the time, t, is registered auto-

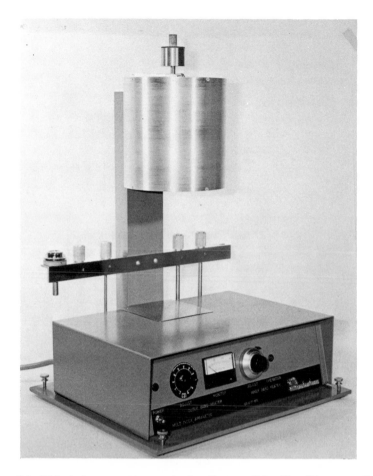

Figure 9-2. Melt index apparatus made by Custom Scientific. (Photo supplied by manufacturer.)

matically, and the melt flow rate in grams/10 minutes is computed and printed.

Figure 9-5 shows a Zwick Capillary Plastometer with an electronic temperature indicator, a manual weight-left device, a plunger guide, a mechanical recorder and a manual extrudate cutting device.

9.3 OTHER WEIGHT-DRIVEN CAPILLARY FLOW TESTERS

ASTM standard test method D569 is based on the Rossi-Peakes flow tester described in Section 4.4.1. A commercial version, the Olsen-

Bakelite Flow Tester, is available from Tinius Olsen. There are two models, both equipped with recorder and bench. One of these uses 150 psi steam for heating, and the other is heated electrically. Prices are in the "G" range.

As pointed out by Black [336], the Rossi-Peakes test has been replaced in many applications by the extrusion plastometer test because of its greater versatility and better control of variables.

Similar to the Rossi-Peakes device is the Koka Flow Tester made by

Figure 9-3. Kayeness model D-7051 extrusion plastometer. (Photo supplied by manufacturer.)

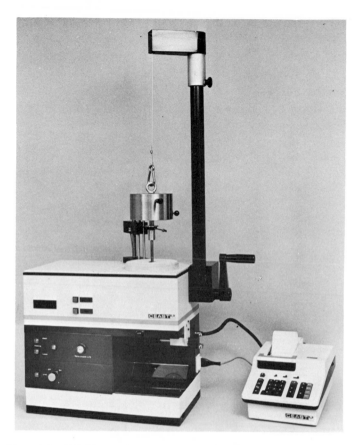

Figure 9-4. CEAST extrusion plastometer. (Photo supplied by manufacturer.)

Shimadzu in Japan. This instrument is somewhat more sophisticated than the Rossi-Peakes, but is designed for the same sort of moldability study in which the melting occurs under load. A built-in recorder plots plunger travel versus time at constant temperature or travel versus temperature during a period of steady temperature increase at a rate of 1°C, 3°C or 6°C/minute. Oyanagi [71] used a modified Koka Flow Tester to carry out a detailed study of extrudate distortion.

9.4 GAS PRESSURE-DRIVEN CAPILLARY RHEOMETERS

The use of compressed air or nitrogen to supply the driving force for polymer flow increases the versatility of a capillary rheometer beyond that of a weight-driven device. The load is continuously variable, and it becomes practical to use higher loads. At the same time, elaborate electronics and automatic control systems are not necessary. However, as explained in Section 4.4.2, constant driving pressure does not necessarily correspond to constant wall shear stress or to constant flow rate.

The simplest type of pneumatic drive involves the direct application of gas pressure to the melt in the barrel, and this is the principle of operation of the CIL viscometer described in Section 4.4.2. A steel sphere, sized to fit snugly in the barrel, is used to ensure the even flow

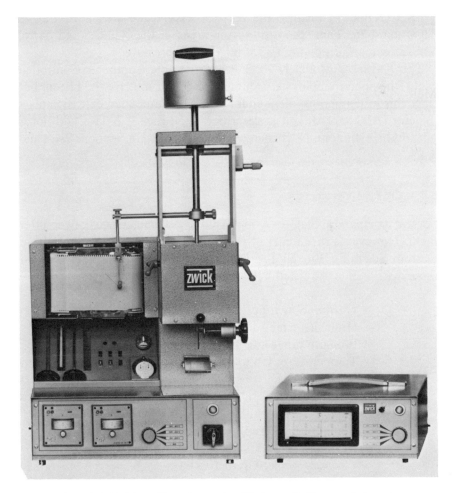

Figure 9-5. Zwick capillary plastometer. (Photo supplied by manufacturer.)

of polymer. Commercial versions of this instrument are described in Section 9.4.1.

The use of a pneumatic cylinder to drive the piston, in place of direct gas pressure, allows the use of higher driving pressures, at the expense of some added complexity of construction. The driving force, F_d, acting on the piston is related to the cylinder area, A_c, and pressure, P_c, according to Equation 9-4.

$$F_d = P_c A_c \qquad (9\text{-}4)$$

This is related, in turn, to the barrel pressure and area according to Equation 4-26. Thus, the ratio of the areas of the cylinder and barrel is, in effect, a multiplying factor for the gas pressure.

The Kayeness Model D-9052 Programmable Computer Controlled Melt Rheometer employs a pneumatic cylinder to drive the piston, but since this instrument is intended for use as an automated extrusion plastometer (ASTM 1238), it is described in Section 9.2. Two additional pneumatic cylinder melt rheometers are described in Sections 9.4.2 and 9.4.3.

9.4.1 CIL Type Viscometers

Several companies that once made CIL viscometers no longer offer them. However, instruments of this type are currently available from Custom Scientific, Burrell and Davenport, and some specifications for these instruments are listed below.

> *Custom Scientific 127HS*
> High shear melt index apparatus
> Pressure ranges: 0–100 psi and 25–650 psi
> Temperature: Up to 400°C
> Price range: F

> *Burrell-Severs A-250 Extrusion Rheometer*
> Manufactured by Burrell
> P_d (max) = 34.5 bars
> Viscosity range: Up to 10 kPa·s
> Temperature: Up to 300°C
> Price range: F

Davenport High Shear Viscometer
Sold in U.S. by Testing Machines, Inc.
Die: $D = 0.5$ mm; $L = 5$ mm
P_d (max) $= 19.6$ MPa
Temperature range: 100–300°C
Price range: F

The Davenport High Shear Viscometer is shown in Figure 9.6.

Figure 9-6. Davenport High Shear Viscometer. (Photo supplied by manufacturer.)

9.4.2 The Sieglaff-McKelvey Capillary Rheometer

Figure 9-7 is a photograph of the Sieglaff-McKelvey Rheometer. The basic specifications of this instrument in its standard configuration are listed below.

Sieglaff-McKelvey Capillary Rheometer
Manufacturer: Tinius Olsen
Marketed by: IMASS, Inc.
Standard dies:
$L = 1$ in.
$D = 0.01, 0.02, 0.04, 0.065, 0.07$ in.
F_p (max) $= 8.9$ kN (0.89 kN available)
$D_b = 7.94$ mm
Temperature: Up to 425°C
Price range: I

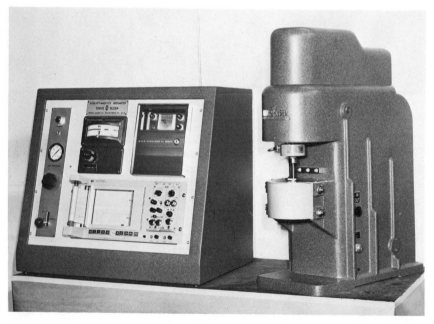

Figure 9-7. Sieglaff-McKelvey Rheometer made by Tinius-Olsen. (Photo supplied by Imass, Inc.)

Clean air or nitrogen at 90 to 100 psia is supplied to a regulator, and the regulated pressure is applied to the pneumatic cylinder. The cylinder piston is linked to the rheometer plunger by means of a lever. With a maximum pressure of 100 psia in the cylinder, the plunger force is 1,500 pounds. The barrel can be removed for cleaning, and its small volume and resulting small filling time make it practical to study easily degraded materials. Wissbrun and Zahorchak [77] have suggested a method for adapting this rheometer to the study of polymers especially sensitive to moisture and oxygen, by provision of a nitrogen blanket.

The force on the plunger is sensed by a load cell* and is recorded on one channel of a high quality, two-channel oscillograph. The piston velocity is recorded on the second channel. There is no pressure or velocity display except on the oscillographic recorder.

Although pneumatic drive instruments are basically constant stress rheometers, the Sieglaff-McKelvey has a feature which permits its use at a controlled rate. For this mode of operation, a hydraulic piston and valve are placed in mechanical parallel with the plunger. The resistance of this system to the motion of the plunger is much greater than that offered by the flow of the melt, so the rate is controlled by the oil flow. Of course, this reduces the force on the piston so that this mode of operation cannot be used when the maximum driving force is desired.

This instrument has been used, for example, in the study of PVC [337], ABS [338], polyolefin blends [339] and polyester [340].

9.4.3 Monsanto Automatic Capillary Rheometer

This pneumatic cylinder rheometer makes use of the barrel-heater unit of the Monsanto extrusion plastometer. Mounted on the same stand as this unit are the air cylinder, regulator and two pressure gauges with ranges of 60 and 160 psig. A regulated, 160 psig supply of clean air or nitrogen is required by the built-in regulator that supplies gas to the cylinder. It is recommended that the supply pressure be at least 20 psi greater than the desired test (cylinder) pressure for good regulation. A chart, based on Equation 9-4, is provided for relating the plunger force,

---

*Lebow Associates.

F_d, to the cylinder pressure. Some additional specifications are tabulated below.

Monsanto Automatic Capillary Rheometer
Standard dies:
  $L = 0.315$ in.; $L/D = 4$
  $L = 0.635$ in.; $L/D = 10$
  $L = 0.615$ in.; $L/D = 15$
Maximum shear stress: 2.0 MPa
Maximum apparent shear rate: 10^5 s^{-1}
Temperature range: 71–315°C
Price range: H

A built-in timer permits the measurement of the time required for the plunger to move through a certain distance. An option called "automatic pressure switching" allows for automatic cycling through four preset pressures in the course of a single run.

An interesting accessory to the Automatic Capillary Rheometer is the Automatic Die Swell Detector. This device can also be used with any capillary extrusion apparatus that allows an unobstructed view of the extrudate. A laser beam is projected through a rotating prism so that it passes across the extrudate. A receiver senses the time during which the sweeping beam is interrupted by the extrudate. A vertical scanning mechanism can be used to measure the shape of the extrudate and thus the short-term time dependency of the swell. Extrudate diameters between 0.25 and 5.0 millimeters can be measured, and the vertical scanning distance is 308 millimeters.

9.4.4 Westover Type High Pressure Rheometer

The capillary rheometer designed by Westover [93] for the study of the effect of elevated pressure on viscosity was described in Section 4.4.2. A commercial model formerly manufactured in the U.S. is no longer available, but the specifications of a version currently manufactured in Japan are given below.

Westover High Pressure Rheometer
Manufacturer: Toyoseiki
Standard die: $D = 2$ mm; $L = 40$ mm

Maximum static pressure: 100 MPa
Maximum driving pressure (difference): 50 MPa
Temperature range: 50–250°C
Price range: I

9.5 CAPILLARY RHEOMETERS WITH HYDRAULIC DRIVE

A servohydraulic system permits the application of large forces precisely controlled on the basis of an error signal. This error signal can be generated by comparing a set point with any of several measured variables, such as plunger speed, plunger force or barrel pressure. Thus, this type of rheometer is equally suitable for studies at a preset rate or at a preset driving pressure. Also, since the power reaches the plunger cylinder via a hydraulic line, there is little restriction on the positioning of the plunger-barrel assembly.

Load control provides a faster response in the control system and is especially well suited for low shear rate studies. However, in this control mode the fluid being tested is itself part of the control loop. Thus, the loop gain is affected by the viscosity and needs to be adjusted to compensate for this. Furthermore, variations in throughput will cause much larger fluctuations in pressure, especially for highly pseudoplastic melts. For these reasons, speed control is generally better for high shear rate studies.

For those wishing to construct their own barrel-heater assembly, general purpose servohydraulic testing machines are available from MTS Systems Corporation and from Instron. A load cell, available from a number of companies, including Lebow and Instron, or a pressure transducer, must also be purchased. The Dynisco transducers are currently popular among melt rheologists.

9.5.1 The Monsanto Processability Tester

This sophisticated test instrument is designed primarily for operation at selected shear rates. A built-in microprocessor allows the operator to select a sequence of rates, and the set point for the hydraulic control system is then switched automatically during the test so that several points on the viscosity curve can be obtained. The rate and driving pressure are shown on a digital readout and recorded on a strip-chart recorder. Basic design parameters are listed below.

Monsanto Processability Tester
Standard dies:
  $L/D = 20$; $D = 0.75, 1.00, 1.50$ mm
  $L/D = 16$; $D = 2.00$ mm
Piston speeds: 0.01–9.99 in./min.
$D_b = 0.75$ in.
Maximum shear stress: 3.0 MPa
Temperature range: 71–232°C (283°C available)
Price range: K

A die swell detector is included, which is based on the same laser system as the Automatic Die Swell Detector described in Section 9.4.3. However, in this instrument, the extrudate is contained in a temperature-controlled chamber so that isothermal swell can be determined.

9.5.2 Other Hydraulic Capillary Rheometers

The Karl Frank High Pressure Capillary Tester (Model 73600SW) can be operated at set speed or at set plunger load. This instrument has a unique feature involving two barrel-heater assemblies which are mounted in such a way that either can be easily rotated into position under the plunger. Thus, one barrel can be cleaned and reloaded while the other is being used in a heat-run cycle. The maximum driving pressure is 2,160 bars.

The Rheograph 2000 is another capillary rheometer that can be operated at constant rate or at constant stress. Specifications are tabulated below.

Rheograph 2000
Manufacturer: Göttfert
U.S. Agent: Automatik U.S.A.
Standard dies: $D = 0.5$ mm; $L = 5, 10, 15$ mm
Piston speeds: 0.008–40 mm/s
P_d (max) $= 2000$ bars (200 MPa)
Choice of three dual-range pressure transducers:
  1. 2,000/1,000 bars
  2. 1,000/500 bars
  3. 200/100 bars
Maximum temperature: 400°C
Price range: K

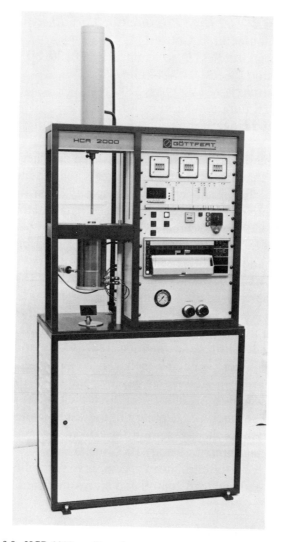

Figure 9-8. HCR 2000 capillary rheometer. (Photo supplied by Göttfert.)

A simpler version of this instrument is the HCR-2000, shown in Figure
9-8. It is designed to accommodate a wide range of Dynisco pressure
transducers. The control system in this rheometer is not of the closed-
loop servo type. One merely turns a valve to regulate the hydraulic pressure rather than presetting the load or speed. Essential data are given
on p. 218.

HCR-2000 High Pressure Capillary Rheometer
Manufacturer: Göttfert
Standard dies: $D = 1$ mm; $L = 10, 20, 30$ mm
Piston speeds: 0.07–44 mm/s
P_d (max) = 2,000 bars (200 MPa)
Maximum temperature: 400°C
Price range: J/K

Also available from Göttfert is a specialized data acquisition system ("Rheodata") that accepts transducer signals and produces printed records of shear rate, shear stress and viscosity. Warmuth [341] has described a more elaborate, computer-based data treatment system.

The use of Göttfert rheometers to study raw rubber mixtures has been described by Wesche [342] and by Toussaint *et al.* [343].

An unusual and interesting servohydraulic capillary rheometer is the Rheoplast RC10 made by Metrilec in Paris. Above the capillary and barrel is a larger preshearing chamber and, above that, an even larger feed chamber. Resin, in the form of pellets or powder, is added to the feed chamber through a hopper. An annular sleeve on the piston can be rotated to speed melting and then lowered to force the melt into the preshearing chamber. Here the melt can be subjected to a specified amount of shearing in an annular gap by rotation of the piston. Then the sleeve is lowered to transfer the sheared melt to the barrel, and the piston forces it through the capillary in the usual manner. A pressure transducer measures the pressure in the barrel. Agassant *et al.* [344] have used this instrument to study PVC melts.

Instron has recently introduced a new servohydraulic capillary rheometer with speed and load control and a pressure transducer mounted in the barrel.

9.6 CAPILLARY RHEOMETERS WITH ELECTROMECHANICAL DRIVE

This type of rheometer is based on a feedback-controlled, constant speed motor. While it is not impossible to design a constant pressure control system, these instruments are best suited for constant rate operation. Depending on the design of the control system, there may be a choice of several prescribed speed multiples, or the speed may be continuously variable. The first such instruments were built in research laboratories for use in conjunction with general purpose, electromechanical testing machines. Now they are available commercially in this form, as well as in the form of a self-contained bench-top device.

9.6.1 The Instron Capillary Rheometers

The Merz-Collwell design, described in Section 4.4.3, has been commercialized by the Instron corporation and is available as an accessory (Model 3210) for use with a standard Instron testing machine (series 1120 or 1190). In these machines, any one of ten plunger speeds can be selected by pushing a button. The compressive load cell measures the total force on the piston rather than the driving pressure in the barrel. The long barrel can accommodate long dies as well as a large volume of polymer. Thus, if polymer degradation is not a problem, a number of shear rates can be used with a single loading of the barrel with resin. Dies with lengths up to 4 inches can be accommodated by the barrel and are available as standard items.

Instron 3210 Capillary Rheometer
Standard dies:
  $D = 0.03, 0.05, 0.06$ in.
      (0.01 in. available)
  $L = 1, 2, 3, 4$ in.
Maximum load $= 20$ kN
A number of load cells are available, including:
  Model 2511-317: 0–0.1 to 0–5 kN in six ranges
  Model 2511-327: 0–0.5 to 0–25 kN in six ranges
Piston speeds: Ten speeds between 0.05 and 500 mm/min.
$D_b = 9.525$ mm
Maximum temperature: 350°C
Price range: G (3210 accessory only)

The Instron Model 3211 is a self-contained capillary rheometer, a photo of which is shown in Figure 9-9. This instrument makes use of the same dies as the 3210 accessory. The barrel has the same diameter but is somewhat shorter. Six speeds can be selected by pushbutton on the standard model. A panel meter indicates the load, and there is a built-in recorder.

Figure 9-9. Instron 3211 capillary rheometer. (Photo supplied by manufacturer.)

Instron 3211 Capillary Rheometer
Same dies as for 3210
F_d ranges: 0.2, 0.5, 1, 2, 5, 10, 20 kN full scale
D_b = 9.525 mm
Piston speeds: 0.6, 2, 6, 20, 60, 200 mm/min.
(others available with optional gear sets)
Maximum temperature: 400°C
Price range: H

Instron Capillary Rheometers have for many years been the principal instruments used for high shear rate studies of molten polymers in North America and are in use in many other countries as well. The use of model 3210 is very attractive for occasional testing in a laboratory that requires a universal testing machine for other mechanical testing. However, if fairly regular use of the capillary rheometer is required, it is much more convenient to have an instrument dedicated to this use, and the model 3211 would seem a better choice in this case.

The 3210 accessory together with a large testing machine makes a more robust unit than the model 3211 self-contained rheometer. However, the longer plunger of the larger unit is more easily bent. The bending of the plunger is a potential problem in all high-load, ram-type capillary rheometers

9.6.2 Other Rheometers with Electromechanical Drive

Toyoseiki of Japan also manufactures a capillary rheometer accessory to be used with its Strograph Universal Tester.

In the Davenport Extrusion Rheometer, shown in Figure 9-10, the plunger is driven by a thyristor-controlled shunt motor acting through a gear train. The speed is continuously variable from 0.125 to 25 centimeters/minute. The driving pressure is monitored by means of a pressure transducer mounted in the wall of the barrel just above the die. The maximum working pressure is 138 MPa.

Davenport Extrusion Rheometer
U.S. agent: Testing Machines, Inc.
Standard die: $D = 1$ mm; $L = 20$ mm
P_d transducer capacity = 21, 34 or 69 MPa
$D_b = 1.91$ cm
Plunger speed range: 0.125–25 cm/min.
Temperature range: 100–300° C
Price range: H

The Rheograph 1000 is a capillary rheometer with drive mechanism powered by a DC motor. This instrument is shown in Figure 9-11. In addition to the standard dies mentioned below, a number of other geometries are available, including annular and slit dies equipped with two wall pressure transducers.

Rheograph 1000
Manufacturer: Göttfert
U.S. agent: Automatik USA
Standard dies:
   $L/D = 10, 20, 30$
   $L = 9.5, 12, 15$ mm

Figure 9-10. Davenport Extrusion Rheometer. (Photo supplied by manufacturer.)

Drive system options:
1. Six speeds from 0.5 to 20 mm/s
2. Six speeds from 0.01 to 0.5 mm/s

P_d transducer capacity = 50, 100, 200, 500, 1000 bars

Maximum temperature: 400°C

Price range: K

9.7 CONTINUOUS CAPILLARY RHEOMETERS

All of the instruments described so far in this chapter are designed for spot testing and laboratory measurement. One starts with a sample of resin and carries out a sequence of steps to determine the desired test result. The time required to obtain the data is typically 10 to 15 minutes, and then substantial time is usually required to clean the rheometer in preparation for running another sample.

For some applications, it is highly desirable to be able to monitor a rheological property of a process stream on a continuous basis, and several commercial instruments are designed for this type of service.

Figure 9-11. Rheograph 1000. (Photo supplied by Göttfert.)

9.7.1 Seiscor Continuous Capillary Rheometers

The current line of Seiscor instruments are developments of the Melt Index Analyzer* and Melt Index Recorder described by Welty [345]. These devices were intended to provide a continuous record of the melt index of a flowing melt. Since the standard melt index test does not lend itself to this type of use, it was necessary to simulate this measurement, in some sense, in a continuous flow device.

Melt under pressure is taken as a side stream from the flow line, conditioned to the desired temperature, and fed to a metering pump driven by a tachometer feedback controlled D.C. motor. The metered flow of melt passes to a cavity (reservoir) that is fitted with a capillary. Cavity pressure is monitored and recorded. If the pressure signal rather than the motor speed is used as the control system input, this rheometer can be made to operate at fixed driving pressure rather than fixed rate.

To simulate the melt index test, the driving pressure is set at 298.2 kPa, which is the nominal driving pressure for the melt index test, a die having the L/D ratio specified in the ASTM standard test method (D1238-73) is used, and the resulting RPM of the metering pump is measured. A number, said to be equivalent to the standard melt index, is then calculated as

$$MI = 10K \cdot S \cdot Q \cdot \rho \, (D_M/D_C)^3 \qquad (9\text{-}5)$$

where:

MI = melt index

Q = pump displacement, cc/rev.

S = RPM of metering pump

K = factor to compensate for piston friction of extrusion plastometer, differences in polymer due to thermal treatment, undefined differences in barrel losses and entrance losses, and errors resulting from the use of a capillary diameter factor based on fully developed Newtonian flow

ρ = melt density

D_M = standard die diameter for melt indexer

D_C = actual capillary diameter used.

---

*U.S. Patent 3,209,581

The use of the cubed diameter ratio to account for the difference between the actual and standard die diameters is based on Equation 4-13, which is strictly valid only for the fully developed flow of a Newtonian fluid.

In practice, the factor K cannot be predicted, and a calibration procedure is used to determine the product $(K \cdot Q \cdot \rho)$ for the resin of interest. Of course, this calibration constant will vary from one resin to another, and with flow rate. Therefore, if the stream being monitored undergoes a change in molecular weight, molecular weight distribution or additive concentration, the calibration constant determined on the basis of the set-point or standard polymer composition will no longer be precisely valid. However, for small changes in composition, Seiscor claims that reliable melt index values are still obtained. In any event, there is no doubt that variations in polymer quality that affect viscosity can be readily detected by this device.

The "AK" series continuous capillary rheometers are available either with console-mounted or panel-mounted controls. In the console unit, shown in Figure 9-12, a dual channel recorder-controller is an integral part of the system, whereas the panel-mounted unit contains no recorder-controller, and the user must provide this.

As explained at the beginning of Section 4.4.1, a melt index value provides a very meager rheological characterization, as it usually involves flow at a rather low shear rate. In recognition of this fact, the Seiscor rheometers are available in a dual cavity configuration with pressure switching, allowing alternate flow at high and low driving pressures. A standard configuration involves the use of the melt index conditions alternated with those of the CIL melt flow test described in Section 4.4.2. A timer can be programmed to switch the flow and pressure at regular intervals so that a nearly continuous record of both flow indexes can be obtained.

For a more complete rheological characterization, Seiscor offers the AK-4100, 4-Point Capillary Rheometer. This instrument is equipped not only with two cavities and pressure switching, but also with a dual-speed pump drive. Automatic switching permits cycling through four preselected shear rates or shear stresses on a continuous basis. This device is designed to accommodate the long Instron dies so that more reliable viscosity data can be obtained. Of course, a single set of four data involving the use, at most, of two L/D ratios is not sufficient for the rigorous determination of the viscosity function by use of the methods

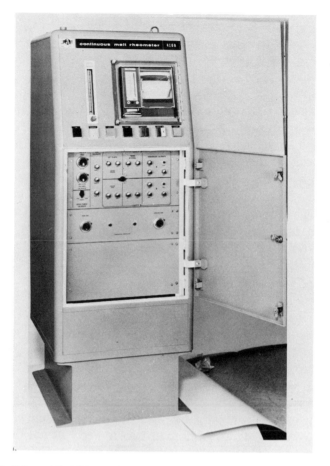

Figure 9-12. Seiscor AK 4100 console model continuous melt rheometer. (Photo supplied by manufacturer.)

of Section 4.3.1. Still, the data do provide an elaborate basis for the detection of variations in polymer quality.

The maximum driving pressure for the Seiscor instruments is 10,000 psi, while the temperature range is 190°C to 300°C. Prices for the Continuous Capillary Rheometer start in the "I" range, while the 4-Point model is in the "K" range.

The basic Seiscor instruments are designed for use with melt streams. However, the AK-5000 Dry Particle Sampling System can be used to melt powder and pellets on a continuous basis and extrude melt into the rheometer. The AK-1000 Polymer Solution Sampler both dries and melts, to provide a feed stream for one of the continuous rheometers.

9.7.2 Göttfert Continuous Capillary Rheometers

Göttfert offers two continuous rheometers, the Kontirheograph and the Side Stream Capillary Rheometer. In both instruments, a metering pump forces melt through a die, and the pressure at the inlet is moni-

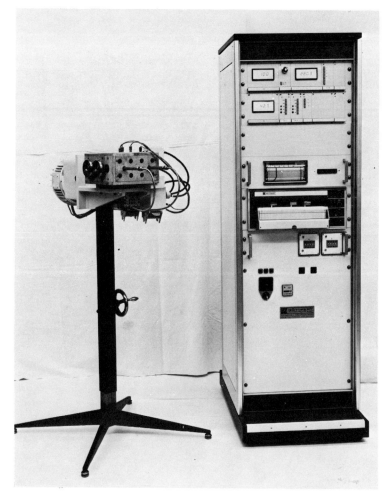

Figure 9-13. Göttfert side stream capillary rheometer. (Photo supplied by manufacturer.)

tored. The standard die has the dimensions specified in the standard melt plastometer test (ASTM 1238), but other designs are available, including a slit die equipped with two wall pressure transducers. The Side Stream Capillary Rheometer, shown in Figure 9-13, is designed to accept melt directly from a process line, while the Kontirheograph shown in Figure 9-14 is a free-standing instrument that includes an extruder so that resin instead of melt can be used as the feed material.

9.8 THE SEISCOR/HAN SLIT RHEOMETER

This sophisticated instrument, shown in Figure 9-15, permits the automatic determination not only of reliable viscosity values but also of approximate values of the first normal stress difference. Like the Seiscor capillary instruments, this device is designed to measure the properties of a side stream of melt coming from a process flow, but in this case, the cavity is fitted with a slit die instead of a capillary. Wall-mounted pressure transducers are located at three positions downstream from the

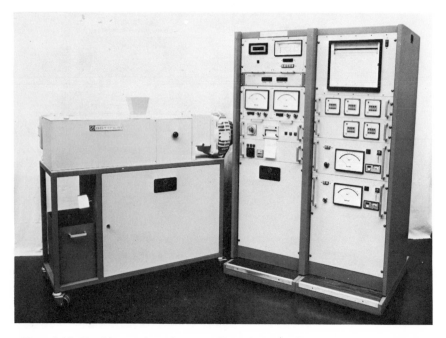

Figure 9-14. Kontirheograph continuous capillary rheometer. (Photo supplied by Göttfert.)

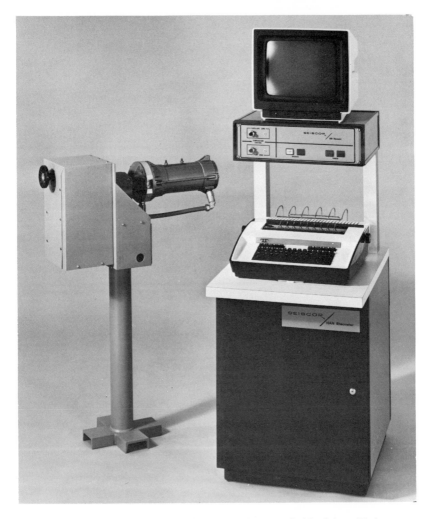

Figure 9-15. Seiscor/Han slit rheometer. (Photo supplied by Seiscor Div.)

entrance. The pump motor is automatically cycled through five speeds so that five sets of wall pressure data can be obtained in one complete cycle.

Measurement parameters such as temperature are set by use of a typewriter. After each complete cycle, a preprogrammed microcomputer analyzes the data. The apparent and actual wall shear rates are computed by use of Equations 4-37 and 4-38, and the wall shear stress is

Table 9-2. Commercial Capillary Rheometers

MANUFACTURER (U.S. AGENT)	MODEL NAME/NUMBER	DRIVE MECHANISM	PRICE RANGE	SECTION IN WHICH DESCRIBED
Burrell	Burrell-Severs A-250	Gas pressure	F	9.4.1
Custom Scientific	CS-127 HS	Gas pressure	F	9.4.1
Davenport (T.M.I.)	High Shear Viscometer	Gas pressure	F	9.4.1
	Extrusion Rheometer	Electromechanical	H	9.6.2
Karl Frank	73600SW High Pressure Capillary Tester	Servohydraulic	K	9.5.2
Göttfert	Rheograph 2000	Servohydraulic	K	9.5.2
Automatik U.S.A.	HCR-2000	Hydraulic	J/K	9.5.2
	Rheograph 1000	Electromechanical	K	9.6.2
	Kontirheograph	Gear pump		9.7.2
	Side Stream Capillary Rheometer	Gear pump		9.7.2
Instron	3210 Accessory	Electromechanical	G	9.6.1
	3211 Capillary Rheometer	Electromechanical	H	9.6.1
Metrilec	Rheoplast RC10	Servohydraulic (with rotation)		9.5.2
Monsanto	Automatic Capillary Rheometer	Gas pressure/ pneumatic cylinder	H	9.4.3
	Processability Tester	Servohydraulic	K	9.5.1
Seiscor	AK Continous Capillary Rheometer	Gear pump	I	9.7.1
	AK-4100 4-Point Capillary Rheometer	Gear pump	K	9.7.1
	Han Slit Rheometer	Gear pump		9.8
Tinius Olsen (Imass, Inc.)	Sieglaff-McKelvey	Gas pressure/pneumatic cylinder	I	9.4.2
Toyoseiki	Westover High Pressure	Gas pressure/hydraulic	I	4.4.2 9.4.4
	Accessory for Strograph	Electromechanical		9.6.2

computed by use of Equation 4-36. This allows the determination of the viscosity. Then the exit pressure is determined for each shear rate, and the first normal stress difference is computed by use of Equation 4-39. The raw data are displayed on a cathode ray tube and the calculated property data are printed.

McGinnis and Han [346] have reported on the use of this instrument to characterize several polyolefins.

The use of exit pressure data to determine the first normal stress difference involves certain uncertainties, which are outlined in Section 4.6. Nevertheless, the Seiscor/Han is the only commercial rheometer now available that is designed to provide normal stress data at high shear rates.

Chapter 10
Commercial Rotational Rheometers

10.1 INTRODUCTION

Rheometric equations and sources of error for rotational rheometers are presented in Chapter 5. These devices are designed to approximate simple shear, and, in the cone plate and concentric cylinder geometries, the shear rate is nearly uniform, certainly far more so than in capillary or slit flow. Thus, rotational rheometers are better suited to the measurement of time-dependent properties and normal stress differences.

On the other hand, for the reasons mentioned in Chapter 5, rotational rheometers are limited to use at rather low shear rates. The shear rate at which secondary flows and free surface distortion become troublesome depends on the detailed geometry and on melt properties, but it is often between 0.1 and 10 s^{-1}. These rates are, of course, quite low compared with those that can occur in extrusion, die flow and injection molding, so that the data obtained cannot be applied in a quantitative way to the modeling of certain aspects of these processes.

Among the many geometries that have been used with rotational rheometers, the most popular has been the cone and plate. Loading and cleaning are relatively simple, and the data can be interpreted directly in terms of the viscometric functions without differentiation. Temperature control is facilitated by the simplicity of the fixtures and is often accomplished simply by enclosing them in an oven with natural or forced convection of the surrounding gaseous medium.

For the measurement of the linear viscoelastic properties, $\eta'(\omega)$ and $G'(\omega)$, the use of eccentric rotating disks was favored for some years after this method was popularized by Professor Bryce Maxwell of Princeton University. A major advantage of this geometry is that the analysis of the experimental data is simple and straightforward, whereas the use of oscillatory shear in the cone and plate or parallel disk geometries requires the determination of the phase shift, δ, reliable values of which were, until recently, difficult to ascertain. On the other hand, use of the ERD mode of operation requires the use of a sophisticated set of transducers which must be at the same time sensitive and noncompliant. Furthermore, the development of inexpensive microprocessors has made it possible to extract precise values for the phase shift from oscillatory shear data. For example, the Dynalyzer* and the Solartron** are digital autocorrelators designed for analyzing data from dynamical mechanical tests of viscoelastic materials. Thus, there has been some tendency to return to the older method for measuring the linear properties.

Three of the instruments described in this chapter can be called "general purpose" rheometers because they can be used to measure both time-dependent viscometric functions and time-dependent properties such as $\eta'(\omega)$ and $G'(\omega)$ with a wide choice of geometries. These are the Weissenberg Rheogoniometer, the Rheometrics Mechanical Spectrometer and the Instron Rotational Rheometer. Versatility is not inexpensive, of course, and these instruments have price tags in the "L" and "M" ranges.

The Rheometer Almighty, made by Iwamoto, is also capable of generating both steady and oscillatory shear, but it is basically a concentric cylinder rheometer.

The other instruments described in this chapter are less versatile and, therefore, less expensive. The Rodem dynamic viscometer, the Orthogonal Rheometer, the Melt Elasticity Tester and the Rheotron, are designed for measuring certain time-dependent material functions, while the devices in Section 10.8 are capable of measuring only viscosity.

For measurement of viscosity at very low shear rates, it is necessary to generate rotational speeds that are at the same time very small and quite stable. This is best accomplished by use of digital position control.

In selecting a rheometer, it is, of course, essential to know the range of stresses and shear rates that are accessible. These depend not only on the basic machine specifications, but also on the particular fixtures used. Clearly, the speed or frequency range and transducer capacities are the basic machine limitations, while geometric factors vary with the choice fixture. The rheometer manufacturers supply a variety of fixtures, but

---

*Made by Dynastatics and marketed in the U.S. by Imass, Inc.
**See Appendix B for the address of the manufacturer.

the user can design and produce any type desired. Thus, it is not possible to specify meaningful ranges for a general purpose machine, and only the basic machine parameters will be given here. The reader can use these to calculate stress and shear rate ranges for particular choices of geometry by using the equations in Chapter 5.

In using cone plate rheometers to study molten polymers, it is necessary to premold disk-shaped samples, and this can often be the most troublesome part of the test procedure. Polymer Products and Services manufactures molds designed especially for this application. Several sizes are available in the "E" price range, the largest one allowing 24 specimen disks to be made at one set of molding conditions. Rheometrics, Inc. also manufactures molds for use with the fixtures that it supplies with its rheometers.

10.2 THE WEISSENBERG RHEOGONIOMETER AND RELATED INSTRUMENTS

The basic features of the Rheogoniometer are described in Section 5.2.3, and it is noted there that commercial versions of this instrument have played a central role in experimental rheology during the past quarter-century. The manufacturer of the Weissenberg Rheogoniometer reports that over 450 are in use in 22 countries, nearly all of these being models 16, 17, or 18.

The current model, designated the R19, is shown in Figure 10-1. It appears to differ from its predecessor primarily in the use of solid state electronics and digital displays. The drive mechanism is entirely mechanical, based on a synchronous motor and a gearbox that allows the selection of 60 speeds from 6×10^{-5} to about 47 radians per second when operated with 60 Hz power. An electromagnetic clutch-brake assembly facilitates stress growth and stress relaxation measurements.

For oscillatory shear, a second motor and drive mechanism is added, and this allows the use of 60 frequencies in the range from 7.6×10^{-5} to 60 Hz at amplitudes between 2 and 30 milliradians.

The drive mechanism is linked to the lower member, which can be a cone, a plate or the outer cylinder of a concentric cylinder arrangement. The standard geometry consists of a cone and a plate, each 5 centimeters in diameter, the cone having an angle of 2°. A number of other sizes and angles are also available.

The upper member is supported by a shaft, which is the rotor of an air bearing. This shaft is, in turn, suspended from a calibrated torsion

Figure 10-1. Weissenberg Rheogoniometer manufactured by Sangamo. (Photo supplied by Sangamo.)

bar, the deflection of which is detected by a displacement transducer capable of reacting to a deflection as small as 10^{-7} radians. Several torsion bars are available for various torque ranges.

Model A, the Rheo-Viscometer, is equipped only for torque measurement, while Model B, the NF Rheo-Visco/Elastometer, can also measure normal thrust on the upper member. In this latter model, the lower member is free to move vertically and is supported by a leaf spring. A positive first normal stress difference tends to force down the lower member and thus deflect the spring. An inductive displacement transducer detects the deflection, and the original gap spacing is restored either manually or by means of a servo system. The spring deflection now provides a measure of the total normal force.

Models A and B are equipped only for steady rotation. Model C, the OS Rheo-Visco/Elastometer, is equipped with a mechanism for sinusoidal oscillation of the lower member and a displacement transducer which produces a signal proportional to the angular position of the lower member and thus to the shear strain. A second drive unit can be added to allow superposed oscillation and steady shear. There is no provision for normal stress measurement.

Model D, the Weissenberg Rheogoniometer, allows the measurement of all the functions mentioned above; i.e., viscosity, first normal stress difference and linear viscoelastic properties. The price of this unit is in the "L" range.*

The limitations of the Rheogoniometer, as reported by many users, together with methods for overcoming these, are described in Section 5.2.3. One problem is that the torque and normal force measurement systems have significant time constants so that reliable stress growth results cannot be obtained. Normal stresses during oscillatory shear are likewise often impossible to measure. To make possible such measurements, Deimos, Ltd. manufactures a piezoelectric sensing head for measuring torque and normal forces in a Weissenberg Rheogoniometer. This device replaces the standard transducers, and the springs, servo system and air bearing are no longer required. The new transducer operates with negligible displacement so that frequency response is much improved. This unit is priced in the "F" range. A liquid cooled model for high temperature studies is also available. The Deimos transducer housing and electronics unit are shown in Figure 10-2.

---

*See Appendix C.

(a)

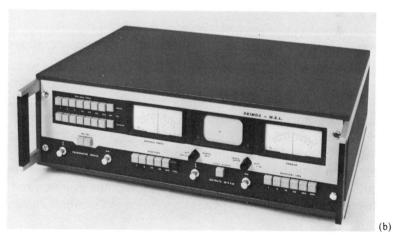

(b)

Figure 10-2. Deimos transducer housing (a) and charge amplifier (b). (Photo supplied by manufacturer.)

10.3 RHEOMETERS MANUFACTURED BY RHEOMETRICS, INC.

In an effort to overcome many of the deficiencies of the rheometers then available, Macosko and Starita [347] designed, around 1970, a general purpose rotational rheometer based on a versatile transducer system capable of measuring torque, normal force and the forces in two directions orthogonal to the axis of rotation. It was thus possible to use the widest possible range of rotational test geometries, including cone plate, parallel disk, concentric cylinder and eccentric rotating disks.

Rheometrics, Inc. was established to manufacture this new device, which was called the Mechanical Spectrometer. The early models were equipped with a tilting mechanism so that the lower test fixture could be set at an angle to the upper fixture, but this feature was eliminated in later models, because it was little used and added significantly to the complexity of the instrument. Its elimination also permitted an improvement in rigidity. The Mechanical Spectrometer is shown in Figure 10-3.

The upper member is mounted on the end of the shaft of a servo-controlled rotary actuator equipped with a high precision bearing. This drive unit can be operated at angular velocities from 10^{-2} to 100 rad/second with an acceleration of over 900 rad/second^2. Alternatively, oscillatory motion can be generated with frequencies from 0.001 to 100 rad/second and amplitudes up to 0.5 radians. Square and triangular waves can also be generated. All wave forms are generated digitally using 16,384 points per cycle.

The lower member is mounted on the force and torque measuring unit. The standard ranges for torque and normal force are 2,000 centimeter-grams* and 2 kilograms (force) full scale. More sensitive transducers with maximum torques of 1, 10 and 100 centimeter-grams are also available. Some of the transducers are available with a temperature control system to minimize drift. An apparatus for calibration is provided that makes use of deadweight loading. Accessories necessary for the use of eccentric rotating disk geometry are available.

The environmental chamber is a forced convection oven capable of operation up to 400°C. The gas in contact with the sample can be air or nitrogen, and if low temperature operation is essential, an optional

---

*To convert from centimeter-grams (force) to meter-newtons, multiply by 9.81×10^{-5}.

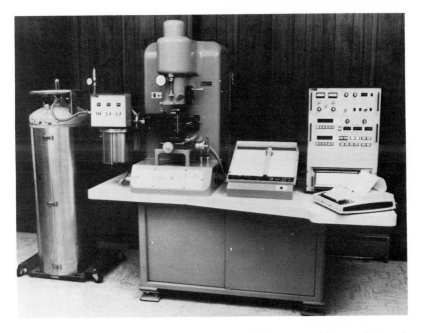

Figure 10-3. Rheometrics Mechanical Spectrometer. (Photo supplied by manufacturer.)

accessory allows the use of liquid nitrogen to produce temperatures down to $-150°C$.

When operating in the frequency mode, the strain and stress signals are sampled and the data processed by a microprocessor. Digital displays, a strip chart recorder, an X-Y recorder and a printer are provided as output media. A keyboard permits the user to set up testing sequences. Automatic sweeps of frequency, temperature and strain are possible, as well as combined frequency and temperature sweeps.

Standard test fixtures consist of two 25 millimeter diameter disks and one cone having an angle of 0.1 radian; a variety of additional test geometries are also available.

Among the many published references to data obtained on a Mechanical Spectrometer are papers on PVC plastisols [348], rigid PVC [349], raw elastomers [350] and polyethylene [351].

The System Four rheometer is a more versatile instrument having all the capabilities of the Mechanical Spectrometer, but also allowing for small amplitude dynamic tests of solid-like materials and rotational flows of low viscosity fluids [352]. A combined analog and digital position control system allows the generation of stable rotation at speeds down to 10^{-3} rad/second. Prices for the Mechanical Spectrometer and the System Four instrument are in the "N" range.

Rheometrics also manufactures two less versatile instruments designed exclusively for dynamic (oscillatory shear) testing. The Dynamic Spectrometer provides for interactive keyboard control of testing and has a frequency range of 0.01 to 500 rad/second (0.0016 to 80 Hertz). Prices are in the "L" range. The Visco-Elastic Tester is a simpler instrument limited to operation at frequencies from 0.1 to 100 rad/second (0.016 to 20 Hertz). It is priced in the "K" range. The Visco-Elastic Tester is shown in Figure 10-4.

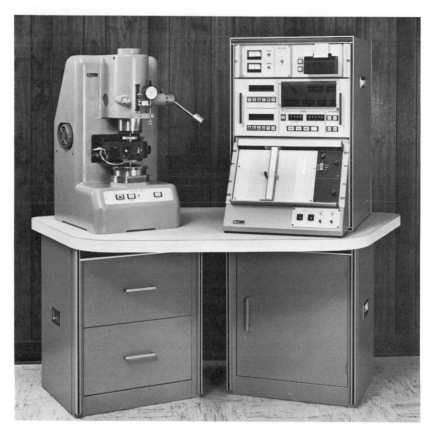

Figure 10-4. Rheometrics Visco-Elastic Tester. (Photo supplied by manufacturer.)

Figure 10-5. Instron Rotational Rheometer. (Photo supplied by manufacturer.)

10.4 THE INSTRON ROTATIONAL RHEOMETER

The Instron Rotational Rheometer was first described by Drislane *et al.* [353].

As can be seen in Figure 10-5, the frame of this instrument is designed to minimize compliance, and the manufacturer specifies that the overall vertical stiffness at the opposing test fixtures is 6×10^6 N/m. The basic instrument, priced in the "L" range, is equipped for measurement of torque and normal force, while options are available to permit oscillatory shear and the use of eccentric rotating disks. The drive motor shaft is supported by an air bearing and is linked directly to the upper rheometer fixture. The servomotor drive is based on an incremental encoder with 128,000 position steps per revolution. This permits the precise setting of speed over the range from 10^{-5} to 2500 RPM (10^{-6} to 250 rad/second). Triangular, square and sinusoidal oscillations in shear rate can be superimposed on steady shear or used by themselves at frequencies from 10^{-2} to 100 Hz.

The basic rheometer, however, cannot be used as it stands for measurement of dynamic viscosity or modulus. For one thing, there is no output strain signal that could be used to determine the phase angle. For dynamic testing, an optional "Analog Angular Position Control" is available. This accessory provides improved position control and a position output signal that can be sent, along with the torque signal, to a frequency response analyzer for determination of the phase shift. A package consisting of a Solartron Analyzer and a Hewlett Packard programmable calculator is available for automatic measurement and evaluation of linear viscoelastic properties.

The Analog Angular Position Control also permits the sudden imposition of strain. In fact, since position is now controlled by a servo loop, any function of time can be used.

The standard torque and normal force transducer has capacities of 2 meter-newtons and 100 newtons, while the high sensitivity transducer measures torques up to 20 millimeter-newtons and normal forces up to 1.0 newton. Six sensitivities may be selected for the load cell amplifier, with the most sensitive selection giving a full scale reading of $\frac{1}{50}$ times the transducer capacity.

The optional thermal chamber is a radiant heat oven capable of sustaining temperatures up to 400°C. A low temperature chamber for tests down to −150°C is also available. An orthogonal force load cell mounted with an air bearing is available for those wishing to use eccentric rotating disks.

10.5 RHEOMETERS MADE BY IWAMOTO SEISAKUSHO

The "Rheometer Almighty" is a concentric cylinder instrument in which the outer cylinder can be rotated at constant speed or oscillated sinusoidally. Steady speeds in the range from 0.015 to 300 RPM are attainable with speed change by means of a stageless reduction gear. A differential transformer detects the deflection of a torsion wire, from which the inner cylinder is suspended, for determination of the torque. A mechanical oscillatory drive unit can generate frequencies between 0.0005 and 10 Hertz with an amplitude of 2 to 3°.

Closely related to the Rheometer Almighty is the 3-D-Rheometer. This latter instrument, however, is capable of generating superposed steady and oscillatory shear and of measuring normal stress differences. The speed and frequency ranges are the same as in the Rheometer Almighty, as is the maximum operating temperature, 300°C, although the maximum temperature is reduced to 150°C when normal stress is being measured. Viscosities of 10^{-2} to 10^5 poise and normal stresses of 0.5 to 100 grams/centimeter^2 can be measured.

10.6 INSTRUMENTS MANUFACTURED BY CUSTOM SCIENTIFIC

Custom Scientific makes two devices developed by Professor Bryce Maxwell of Princeton University to measure viscoelastic properties of melts. The Orthogonal Rheometer was the first commercial instrument making use of the eccentric rotating disk geometry. The speed range is 0.0035 to 175 RPM. The maximum X and Y displacement of the disk axes is 0.40 inch while the maximum gap is 0.50 inch. A clutch and brake assembly makes possible transient tests such as stress relaxation.

The Melt Elasticity Tester is a concentric cylinder instrument designed expressly for transient shear tests, including stress growth and relaxation and recoverable strain. It is designed to minimize complexity and thus cost. The outer cylinder is driven by a variable speed drive, while a lever arm transmits the resulting torque on the inner cylinder to a load cell. For stress relaxation, a rigid stop acts as a brake. For the measurement of recoverable strain, the lever arm is suddenly disconnected from the load cell by means of a solenoid, and the angular position of the inner cylinder is followed by means of a light beam and recorded photographically. By increasing the frequency of exposures in the early stages of recovery, the details of the rapid initial recovery can be observed.

The ratio of gap width to radius is about ⅛, so the maintenance of uniform strain and temperature is far from ideal. However, the Melt Elasticity Tester is one of the very few commercial instruments permitting the direct measurement of recoverable strain. Also, compared with the general purpose rheometers described earlier in this chapter, its price, in the "G" range, is relatively modest. Maxwell and Nguyen [142] used a Melt Elasticity Tester to study the effect of molecular weight on recoverable strain.

10.7 THE BRABENDER RHEOTRON

The Rheotron is basically a concentric cylinder rheometer in which the outer cylinder is rotated and the torque on the inner cylinder is measured. However, it is possible to substitute a shallow cup for the outer cylinder and a cone for the inner cylinder so that cone plate flow can be generated. A normal force sensor is available for use with this geometry. Speeds in the range of 0.05 to 1000 RPM are available, and a magnetic coupling allows the carrying out of stress growth experiments. A brake prevents recoil so that stress relaxation can also be determined. The Rheotron is shown in Figure 10-6.

Torque measurement is accomplished by use of a lever arm acting against a spring, with the deflection detected by an inductive linear displacement transducer. Thermostatic liquid is circulated through a chamber surrounding the outer cylinder, but there is no temperature control for the inner cylinder. Heat flowing between the fluid under test and the thermostatting fluid must pass through two steel cylinder walls and a dry bearing bushing. When the cone plate geometry is used, the situation is less satisfactory. Thus, temperature control is only fair in this rheometer. On the other hand, its price is modest, in the "H" range.

Heinz and Rothenpieler [354] used the Rheotron to study molten polyethylene, and Seeger and Heinz [355] have used a Rheotron to measure the second normal stress difference by means of the extended cone and plate flow described in Section 5.2.5.

10.8 THE RODEM DYNAMIC VISCOMETER

The Rodem instrument is based on a design developed by K. te Nijenhuis at the Technische Hogeschool, Delft. It is an improved version of the apparatus described earlier by den Otter [144]. The inner cylinder is suspended between two vertical torsion wires, and the upper end of the upper wire is subjected to torsional oscillation by means of a crank mechanism. This angular displacement, as well as that of the cylinder, is monitored by means of light beams reflected from mirrors mounted

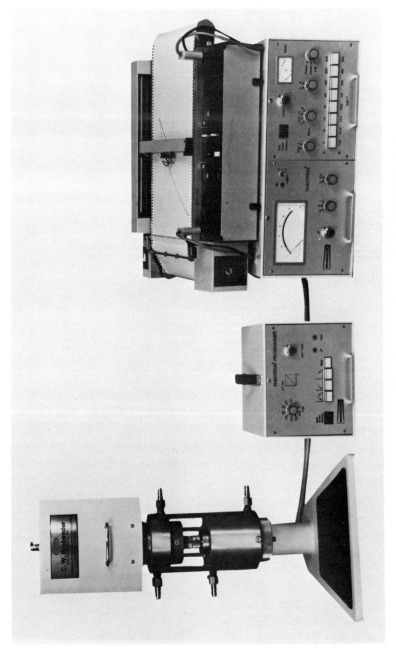

Figure 10-6. Rheotron rotational rheometer. (Photo supplied by C. W. Brabender.)

on the clamps holding the upper and lower ends of the upper torsion wire.

Standard cylinders allow the use of gap widths from 0.5 to 5 millimeters. Angular amplitudes up to 35° are possible, and the frequency can be varied continuously over two decades. The use of interchangeable gear boxes permits dynamic measurements at frequencies from 5 × 10^{-5} to 100 Hz. Temperatures can be controlled at levels between −40°C and 250°C, and digital readout of data is provided. The price is in the "L" range.

10.9 MELT VISCOMETERS

The instruments in this section are designed only for approximate measurement of viscosity. They have the advantages of simplicity and relatively low cost, but no viscoelastic (time-dependent) properties can be determined, and the uniformity of temperature and shear rate are, in general, poor.

10.9.1 The Davenport Cone and Plate Viscometer

The instrument shown in Figure 10-7 is an admirably simple cone and plate viscometer in which the torque is applied by means of a weight suspended from a string. The angular displacement is detected by means of a light beam reflected from a mirror on the cone shaft to a scale. An electric heater allows operation up to 190°C, but the initial heating time is 45 minutes or more at this temperature.

With some simple modifications, this instrument could be adapted to the measurement of creep and creep recovery functions.

10.9.2 Mooney Shearing Disc Viscometers

The basic features of the Mooney viscometer were described in Section 5.4. Commercial versions of this instrument are available from three manufacturers, as indicated below.

Monsanto Standard Mooney
Karl Frank Model 76520
Scott Tester STI/200

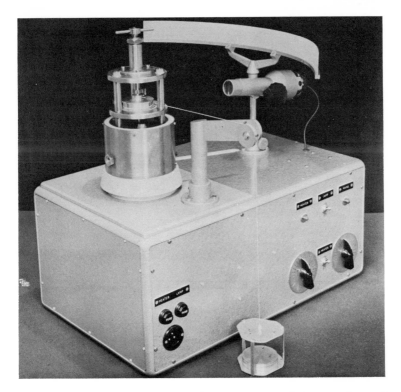

Figure 10-7. Davenport cone and plate viscometer. (Photo supplied by manufacturer.)

The Scott version is marketed by GCA/Precision Scientific Co. and is priced in the "H" range. It is shown in Figure 10-8. Various accessories are available on all three models to provide various degrees of automation either in operation or the analysis of data.

10.9.3 Brookfield Thermosel System

This is an accessory for the standard Brookfield line of viscometers that permits the heating of the sample to temperatures as high as 260°C. It is designed for adhesives,* but with the HB series of viscometers, viscosities up to 1.6×10^4 Pa·s can be measured. Thus, this device might be of use with some lower viscosity molten plastics. The inner cylinder

---

*ASTM Test D 3236-73 refers to this viscometer.

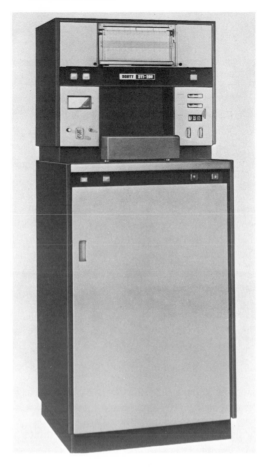

Figure 10-8. Mooney viscometer; Scott Tester STI/200. (Photo supplied by GCA/Precision Scientific.)

is driven by a synchronous motor operating through a gear train. The torque is sensed by a spring and indicated by a needle deflection on a dial that rotates with the inner cylinder. Price for a complete system is in the "C" range.

10.9.4 The Iwamoto Seisakusho Viscometers

The "Melt Rheometer" is a robust, concentric cylinder rheometer capable of measuring viscosities in the range of 5.0 to 5×10^4 Pa·s at pressures up to about 2 MPa and temperatures as high as 300°C. The outer cylinder can be rotated at seven speeds between 0.8 and 100 RPM.

The "Mini-Viscometer" is a similar instrument designed for use when sample size is limited or when rapid testing is desirable to prevent degradation. Viscosities between 10^{-3} and 10^4 Pa·s at pressures up to about 1 MPa can be measured.

Chapter 11
Other Commercial Melt
Rheometers and Testers

The instruments described in Chapters 9 and 10 are useful only for the study of properties relevant to shear flows. The devices mentioned in this chapter are designed to generate flows that are either primarily extensional in nature or combinations of shear and extension. Thus, they can often provide information different from that obtainable by use of a shear rheometer.

11.1 MELT STRENGTH TESTERS

A tester based on the design of Meissner (see Section 7.7.1) is the Rheotens, made by Göttfert. This device consists of a pair of gears, making up a rotary clamp, mounted on the end of one arm of a lever. The other arm of the lever is supported by a spring, and its displacement is sensed by a transducer producing a signal proportional to the force exerted on the clamp by the extrudate filament.

The Rheotens system consists of two units. One contains the rotary clamp, the force transducer and a tachometer feedback motor to drive the clamp. The second unit contains the motor speed control system and a control panel for setting the motor speed. An X-Y recorder can be used to produce a plot of force versus speed.

In order to generate the molten filament, auxiliary apparatus is required which provides for downward extrusion through a die with sufficient clearance under the die to allow for the positioning of the rotary clamp unit. The Minex unit mentioned in Section 9.2 is one possible such device, while Göttfert also offers a special crosshead die to adapt its Extrusiometer or Torsiograph units to this purpose. Another possibility is to use a capillary rheometer to generate the filament.

Similar to the Rheotens is the Melt Tension Tester made by Toyoseiki. It is designed for use with a melt indexer or a specially designed, fixed speed vertical extruder.

Toyoseiki also offers a "Melt Strength Device" as an accessory for use with its Labo Plastomil torque rheometer, although it could be used with any extrusion device producing a horizontal filament. The cooled filament passes over a series of three pulleys such that the central pulley experiences an upward force equal to the tension in the filament. This pulley is mounted on a load cell. After passing back upwards over the third pulley, the filament passes between the motor-driven take-up rolls. The range of linear speeds is from 3 to 78 meters/minute, while the standard load cell has a capacity of 20 grams.

11.2 EXTENSIONAL RHEOMETERS

Two commercial rheometers are available that permit the study of the uniaxial extension of molten polymers. The Rheometrics Extensional Rheometer and the Göttfert Rheostrain are versions of the instrument developed by Münstedt [247] and are built under license from BASF. They are in the "K" price range.

The Münstedt instrument, described in section 6.1.5, operates by stretching a sample held in a vertical position in an oil bath, with the density of the oil matched to that of the melt.

Temperature uniformity is essential for reliable results, and the oil bath is carefully designed to minimize temperature gradients. It consists of a tall, cylindrical, jacketed glass container. Heat is supplied by oil circulating in the jacket, and two additional glass cylinders surround the jacketed vessel to reduce heat losses to the room. The entire bath assembly can be lowered to expose the force transducer and permit the positioning of samples. The maximum operating temperature is 250°C.

The sample, typically 5 millimeters in diameter and 2 centimeters long, has a small metal fixture glued to each end. One end fixture is linked to a force transducer mounted at the bottom of the bath, on the end of a long rod.

The fixture at the upper end of the specimen is hooked to the end of a rod or flexible tape whose vertical position is controlled by a servo drive system. Signals proportional to both force and length are available so that the stress and strain can be computed according to Equations 6-8 and 6-3 if the melt is assumed incompressible. With appropriate circuitry, the servo drive system can be controlled in such a way as to generate either a constant strain rate (stress growth) or a constant stress (creep) deformation. In addition, stress relaxation after cessation of straining can be measured, and in the Rheometrics instrument creep recovery can be monitored.

A special vacuum mold for preparing up to 70 specimens at a time for use in one of the extensional rheometers described here is available from Polymer Products and Services at a price in the "D" range. A more elaborate sample preparation unit consisting of a vacuum mold for 15 samples, a cutting machine, an etching fixture and a gluing fixture is available from Rheometrics in the "G" price range.

In the Rheometrics Extensional Rheometer, shown in Figure 11-1, the upper clamp is mounted on the end of a rod whose position is controlled by a servomotor. Strain rates between 10^{-3} and 5 s^{-1} are accessible, and the ranges for the force transducer are 0.1, 1.0 and 10 newtons. By means of a servo loop, creep tests can be carried out at constant stresses between 10^3 and 10^6 Pa, and recoil can be measured. The max-

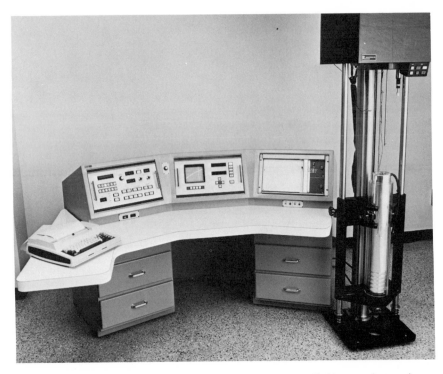

Figure 11-1. Rheometrics Extensional Rheometer. (Photo supplied by manufacturer.)

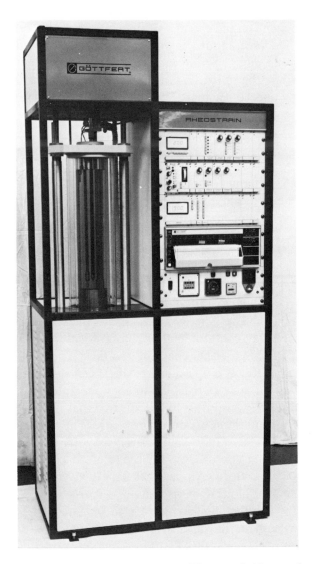

Figure 11.2. Göttfert Rheostrain extensiometer. (Photo supplied by manufacturer.)

imum sample extension is 600 millimeters, corresponding to a Hencky strain of about 4. This instrument has been used by Lobe and Macosko [356] to evaluate formulations for use in the Xerographic reproduction process.

In the Göttfert Rheostrain, shown in Figure 11-2, the upper sample clamp is suspended on the end of a flexible tape, as in the original Mün-stedt design. The tape, in turn, is wound onto a wheel turned by a servo-controlled torque motor. Strain rates of 0.005, 0.05, 0.5 and 2 s^{-1} can be selected, while the full scale ranges for the force transducer are 0.1, 0.2, 0.4, 1, 2, 4 and 10 newtons. The maximum extended sample length is 500 millimeters, giving a maximum Hencky strain of about 3.9.

11.3 TESTERS BASED ON SQUEEZE FLOWS

In Section 7.5, equations for interpreting the results of squeeze flow tests were presented, and it was concluded that such tests may be useful for the classification of materials according to their low-shear-rate vis-cous behavior. The required apparatus is quite simple, involving a fixed lower platen and a weight-loaded upper platen whose position can be indicated as a function of time.

The Parallel Plate Rheometer is an accessory for the Thermome-chanical Analyzer made by the Instrument Products Division of the DuPont Company. The Thermomechanical Analyzer is designed to carry out a variety of constant-load or constant-length mechanical tests at a temperature that is either constant or a linear function of time. An LVDT provides a signal proportional to displacement, and the temper-ature can be varied at rates from 0.5 to 50°C/minute. At high heating and cooling rates, however, it is to be expected that the sample temper-ature will lag behind the set-point temperature. The temperature range is −150°C to 800°C. The manufacturer suggests that this device is useful for measuring viscosities in the range of 10 to 10^7 Pa·s at shear rates from 10^{-5} to 1.0 s^{-1}. An optional data analyzer provides for on-line calculation and plotting of viscosity.

Whereas the DuPont device appears to be intended for use as a squeeze flow rheometer in which the effective sample area remains con-stant, the Tinius Olsen Parallel Plate Plastometer is designed for use in the constant sample volume mode, as specified in ASTM standard test method D-926. In this device, the assembly holding the plates is placed in an oven. The upper platen is fastened to the end of a shaft that passes through a hole in the top of the oven. A load of up to 60 kilograms is applied to this shaft by means of weights suspended from the end of a lever arm. A dial indicator shows the distance between the plates. The manufacturer recommends this device for the measurement of viscosities in the range of 10^3 to 10^8 Pa·s at low shear rates. The price is in the "G" range.

The Consistometer, made by Haake in Germany, and available in North America from Haake, Inc., is a multipurpose, weight-loaded tester for the study of very viscous materials at temperatures up to 350°C. One of the standard test fixtures allows it to be used to generate squeeze film flow. This is a simple device, and the manufacturer suggests its accuracy is about 15%.

11.4 TORQUE RHEOMETERS

The basic features of a torque rheometer are described in Section 7.3, where it is concluded that such a device is actually a tester and cannot be used to determine reliable values for viscosity or any other well-defined material property. On the other hand, hundreds of these instruments are currently used in industrial applications ranging from product development to quality control. Originally introduced in Germany over 50 years ago for use with flour doughs, torque rheometers are now widely used in the plastics industry in connection with the compounding of PVC and rubber and the processing of thermoplastics.

The basic components are a motor, a torque sensor and a mixing head. The motor may have either a fixed speed or a variable speed drive, while the torque sensor may be mechanical or electronic. Motors are usually SCR-controlled and are available in sizes ranging from 1 to 15 horsepower. Motors of at least 3 horsepower are usually selected for use in the plastics industry. Torques on the heaviest units can reach 60 meter-kilograms.

Since the power requirement is equal to the product of torque and speed, as shown in Equation 11-1, the maximum torque that can be generated by a given motor is inversely proportional to the speed.

$$\text{Power } (W) = M \, (\text{m} \cdot \text{N}) \times \Omega \, (\text{s}^{-1}) \qquad (11\text{-}1a)$$

or

$$\text{Power (horsepower)}$$
$$= (1.38 \times 10^{-3}) \times M \, (\text{m} \cdot \text{kg}) \times S \, (\text{RPM}) \qquad (11\text{-}1b)$$

Torque ranges are often specified in meter-grams (m·g) or meter-kilograms (m·kg). To convert to meter-newtons (m·N), which are the correct S.I. units, the following factors can be used.

$$M \text{ (m·N)} = M \text{ (m·g)} \times 0.00981$$
$$M \text{ (m·N)} = M \text{ (m·kg)} \times 9.81$$

The "mixing head" or "mixing sensor" consists of a steel cell in which the sample fills a "bowl" consisting of two adjoining cylindrical chambers. This cell may be heated either by a circulating thermal fluid or by electrical heating elements. Electrical heating is more convenient and allows faster initial heating, but it does not generate as uniform a temperature field. Once mixing is begun, considerable dissipative heat can be generated, and it becomes necessary to remove heat. If electrical heaters are used, this is usually accomplished by air circulation. Thermal fluid heating can be used up to temperatures of 300°C, but between 250°C and 400°C, electrical heating is preferable. Thermocouples are often mounted in the mixing head to measure the temperature of the melt and the metal wall. A common test procedure is to increase the temperature at a linear rate and measure the effect on the torque.

Rotating in the cylindrical cavities are a set of matched rotors or "blades." One of these is coupled directly to the motor drive, while the second is driven by a gear mechanism at a speed that usually differs from that of the motor. A variety of rotor shapes is available, including sigma, roller, cam and Banbury types. The deformation pattern is a complex one involving alternate shearing and relaxation. The roller-type blades produce a relatively high shear rate, and are often used with thermoplastics. Sigma blades were originally used to test flour doughs and are now used to study dry powder blending of PVC and other materials. They produce a lower shear rate than other types of blade. The cam produces medium shear rate deformation and is often used to test elastometers. The Banbury blades are designed to simulate the action of commercial mixers used for rubber and plastics compounding.

Sample volume in the mixing head ranges from 30 to 650 milliliters. Mixing heads are sold either with integral rotors or in a configuration that allows the use of interchangeable rotors. Also available from some manufacturers are devices for nitrogen purging and for applying pressure to a loading ram, either by means of a weight or an air cylinder.

Most torque rheometer manufacturers offer a variety of extruders that can be coupled to the motor in place of the mixing head. When equipped with a processing die, various polymer conversion operations can be simulated in the laboratory, including profile extrusion and film blowing. If a capillary or slit die is used, and pressure transducers are installed at the appropriate locations, reliable values of viscosity can be measured. However, since it is the screw speed rather than the throughput that is controlled, the flow rate must be determined by collecting and weighing extrudate. Also, since there is likely to be significant viscous heating in the extruder, temperature control is not as straightforward as in the case of piston or pressure-driven capillary viscometers.

An early application of torque rheometers was the absorption of plasticizer by dry blend PVC. ASTM standard test method D2396 for plasticizer absorption is based on the use of sigma-type blades at 63 RPM and 88°C. Arendt [357] used a torque rheometer to study both dry blends and PVC resins, employing a constant rate of temperature increase. ASTM standard test method D2538, for the fusion rate of PVC, involves the use of roller-type blades. Petrich [358] evaluated PVC lubricants, while Collins et al. [359] studied the thermal stability of PVC compounds. Patterson and Noah [360] used a torque rheometer in their study of the dispersion of polymer blends, and Allen and Williams [361] determined the molding temperature of several thermoplastics by increasing the temperature in a roller mixer.

Goodrich [362] used a torque rheometer to characterize polypropylene. He used an oil-circulation temperature control system and found that it was impossible to keep the temperature at a preset level. He suggested a technique for reducing all his data to a common reference temperature. Willard [363] compared the effectiveness of a torque rheometer with that of a Mechanical Spectrometer equipped with eccentric rotating disks, for the evaluation of glass-filled molding compounds. He found the linear viscoelasticity technique superior in several ways. This is not very surprising, however, because the Mechanical Spectrometer is a very sophisticated and complex instrument costing many times as much as a basic torque rheometer.

Extensive bibliographies of torque rheometer applications are available from C. W. Brabender Instruments, Inc. [364] and from Haake, Inc. [365].

Prices for a basic torque rheometer are in the "G" range.

Before describing the equipment available from various manufacturers, it is perhaps useful to give a brief summary of the origins of the two major suppliers: C. W. Brabender, Inc. and Haake, Inc. Mr. C. W. Brabender started making torque rheometers for use in flour testing over 50 years ago in Germany, and he established an American affiliate called the Brabender Corporation. At the outbreak of World War II, the American company severed its ties with the German parent. After the war, C. W. Brabender started a new U.S. company called C. W. Brabender Instruments, Inc. For a number of years, the Brabender Corporation and C. W. Brabender Instruments, Inc. marketed torque rheometers nearly identical in appearance. The Brabender Corporation was later purchased by Haake, Inc.

11.4.1 C. W. Brabender Plasti-corder

The basic Plasti-corder models are listed in Table 11-1. The "PL" model employs a mechanical torque indicator, while the "EPL" models are based on electronic torque measurement.

Over a dozen types of mixer head are available, including all the designs and options mentioned in the introduction to Section 11.4. Both ¾ inch and 1¼ inch extruders are also available together with processing and rheometric dies. Figure 11-3 shows a PL model Plasti-corder equipped with an extruder and a rheometric capillary die.

11.4.2 Haake Rheocord

The Haake Rheocord is available in a range of models as shown in Table 11-2. The "M" models use a mechanical torque sensor, while the

Table 11-1   Plasti-corder Models (C. W. Brabender Instruments, Inc.)

MODEL NUMBER	MOTOR SIZE (HORSEPOWER)	MAXIMUM SPEED (RPM)	MEASURABLE TORQUE (m·kg)	NUMBER OF TORQUE RANGES
PL-V300	3.0	200	14 (max)	6
EPL-V3301	3.0	90	1–20	5
EPL-V3302	3.0	150	0.2–10	6
EPL-V5501	5.0	90	1–40	5
EPL-V5502	5.0	150	1–20	5
EPL-V7752	7.5	90	1–60	5

Figure 11-3. Model PL Plasti-corder with extruder head and capillary die. (Photo supplied by C. W. Brabender.)

Table 11-2 Haake Rheocord Models

MODEL NUMBER	MOTOR SIZE (HORSEPOWER)	MAXIMUM SPEED (RPM)	MAXIMUM MEASURABLE TORQUE (m·kg)	NUMBER OF TORQUE RANGES
M-3000	3.0	150	15	8
M-5000	5.0	250	15	8
EU-3V	3	150	15	3
EU-5V	5	250	15	3
EU-10V	10	250	29	4
EU-150V	15	250	45	4

"E" models have electronic torque meters. The M-5000 has been the most popular model in the past, while the "EU" models represent this company's most advanced technology.

Sixteen varieties of mixing head are available in the "Rheomix" line. Also offered are the "Rheomex" line of ¾ inch and 1 inch extruders and the "Rheolex" line of 300 millimeter extruders. The "Transfermix" is a modular version of the 30 millimeter extruder. A wide assortment of processing and rheometrical dies are available for use with these extruders.

11.4.3 Göttfert Torsiograph

This torque rheometer employs a thyristor-controlled motor with a speed range of 1 to 140 RPM and an electronic torque sensor. Eight torque ranges can be selected with full scale readings from 5 to 200 m·N. The electrically heated mixing head, capable of operating at 400°C, has "triangular" blades, which appear to be similar to the "roller" type. Several 20 millimeter extruders are also available. The "Extrusiometer" is a highly instrumented, heavy-duty, 30 millimeter extruder. Figure 11-4 is a photo of the Göttfert Torsiograph.

Figure 11-4. Göttfert Torsiograph. (Photo supplied by manufacturer.)

Table 11-3 Labo-Plastomill Models

MODEL NUMBER	MOTOR SIZE (HORSEPOWER)	MAXIMUM SPEED (RPM)	MAXIMUM MEASURABLE TORQUE (m·kg)	NUMBER OF TORQUE RANGES
10-230	3	230	10	6
15-150	3	150	15	6
15-230	5	230	15	6
25-150	5	150	25	6
40-100	5	100	40	6

11.4.4 Toyoseiki Labo Plastomill

The Toyoseiki torque rheometer models listed in Table 11-3 are available as either floor or table types. Torque sensing is electronic.

Eight mixer heads and four extruders are available. All but one of these have electrical heating.

11.4.5 Hampden/RAPRA Variable Torque Rheometer

This instrument was developed by the Rubber and Plastics Research Association of Great Britain [366]. A fixed speed version can be operated at 30 or 60 RPM, while a variable speed version can be operated between 10 and 100 RPM. The torque is monitored electronically, and the maximum torque is 16 m·kg. The mixing head is heated by circulating oil and is capable of operation from 20°C to 200°C. A pneumatic cylinder-operated ram is provided for loading.

Chapter 12
Epilogue

As a result of the organizational plan adopted by the author for this book, several matters that are probably worth mentioning do not seem to fit neatly into the main outline and have been lumped together in this final chapter. This potpourri includes a look at the deficiencies in our present repertory of tests and instruments, which pose challenging problems for today's rheometer designers.

12.1 MEASUREMENT OF CROSSLINKING SYSTEMS

Crosslinked systems include thermosetting plastics and cured rubbers. Thermosetting plastics are often formed from low viscosity liquids, while raw rubber is a very viscous, tough material, which is not processed in the same way as molten thermoplastics. Nevertheless, these polymeric systems are of great commercial importance, and rheometrical techniques can provide valuable information about the crosslinking process.

The curing of thermosetting systems and crosslinkable polymers, such as polyesters, has often been studied by means of rheometrical techniques. Of course, special problems arise with these materials. Temperature control is more difficult because of the heat of reaction and because of the effect of temperature on the reaction rate. Also, once crosslinking has proceeded to a significant extent, removal of the sample from the rheometer can be quite difficult.

The traditional tester in this field is the spiral flow mold used, for example, by Heinle and Rodgers [367]. However, more meaningful results can be obtained by measuring viscosity as a function of time at a constant temperature and strain rate. Because of the relative ease of cleaning, cone and plate fixtures have often been used [368, 369, 370] to measure viscosity and dynamic properties as functions of time during the crosslinking process. Richter and Macosko [371] designed a special concentric cylinder fixture for use in a Mechanical Spectrometer to study a RIM-type system. The glass fixture components are disposable so that cleaning is not necessary.

Another rheometer having disposable components is the capillary instrument of Shaw *et al.* [372], which is a modification of the Instron Capillary Rheometer involving a disposable, thin-walled insert made from stainless steel tubing.

The curing of rubbers is often characterized by means of oscillatory shear, and several commercial rheometers have been designed especially for this application. For example, Monsanto makes several cure-meters based on oscillating disks or cones.*

12.2 LONG-TERM HISTORY EFFECTS

It has been observed that the effects of a period of intense shearing can affect the subsequent mechanical response of a melt for very long periods of time. To study this phenomenon, one can devise special test procedures making use of standard rheometers. For example, Dealy and Tsang [373] carried out "interrupted shear" and "reduction in shear rate" tests using a Mechanical Spectrometer. However, rotational rheometers are limited to use at rather low shear rates, and there has, therefore, been some incentive to develop special capillary rheometers allowing uniform shearing prior to extrusion. Hanson [98] developed a rheometer in which the melt can be sheared between concentric cylinders prior to extrusion through an orifice, and the Rheoplast, described in Section 9.5.2, is a commercial rheometer based on a similar principle.

Much remains to be learned about long-term history effects, but these could be of great practical importance, especially in situations where melt is subjected to consecutive shearing processes at decreasing rates of shear. This occurs, for example, when a melt is extruded from a die and then subjected to a subsequent forming process. Indeed, Maxwell and Plumeri [374] found that the viscoelastic response of melt formed by heating pellets is strongly influenced by the extrusion rate involved in the production of the pellets.

---

*Models 100 and 750 and the "Cone Rheometer."

12.3   NEEDED: AN ACCURATE COMMERCIAL CREEPMETER

Several experimental creepmeters are mentioned in Chapter 5, but only one commercial rheometer is designed for use as a creepmeter. This is the Melt Elasticity Tester described in Section 10.6 and used, for example, by Nguyen and Maxwell [375]. This device, while admirable for its simplicity and relatively modest cost, is not a scientific instrument because of its large gap and unsophisticated temperature control system.

Two commercial cone plate rheometers that can be used for constant stress experiments are the Deer Rheometer (Rheometer Marketing, Limited) and the Visco-Elastic Analyzer (Integrated Petronic Systems, Ltd.). However, these instruments are not designed for use at melt temperatures with highly viscous systems. The Davenport viscometer described in Section 10.9.1 could be modified to carry out creep tests, but it seems a shame none of the more elaborate rotational melt rheometers can be used for this type of test.

12.4   WANTED: ELASTIC PROPERTIES AT HIGH SHEAR RATES

One of the most serious deficiencies in our current repertory of tests and instruments is our inability to study viscoelasticity at high shear rates. Rotational rheometers are limited to use at low shear rates, and capillary and slit flows are not suitable for transient testing because of the nonuniform shear rate. It has been suggested that the measurement of exit pressure or hole pressure makes it possible to obtain information about normal stress differences at high shear rates, but a sound theoretical basis for these measurements is lacking, so that the use of such tests is limited to on-line quality control applications.

To avoid the limits of rotational flow and at the same time provide a uniform shear field, the only alternative would seem to be plane Couette flow, as in a parallel plate rheometer, but this type of rheometer has the disadvantages of limited total strain and large end effects. This represents the greatest challenge facing today's rheometer designers.

12.5   NEEDED: A SIMPLE INEXPENSIVE ELASTICITY TESTER

In reading Chapters 9 through 11, the reader has probably noted that nearly all of the recently introduced commercial rheometers are highly sophisticated instruments that are, justifiably, costly. The "melt indexer" or extrusion plastometer remains after many years the only low cost, simple melt tester. It is, however, very limited in its capabilities and gives no information about viscoelastic response. It seems likely that there is a substantial market for a simple elasticity tester for melts. The Melt Elasticity Tester (Section 10.6) is a step in this direction.

References

1. R. B. Bird, R. C. Armstrong and O. Hassager, *Dynamics of Polymeric Liquids: Volume I, Fluid Mechanics.* John Wiley & Sons, NY (1977).
2. J. D. Ferry, *Viscoelastic Properties of Polymers,* 3rd Ed. John Wiley & Sons, NY (1980).
3. W. W. Graessley, "The Entanglement Concept in Polymer Rheology," *Adv. Polym. Sci.* **16** (1974).
4. G. V. Vinogradov and A. Ya. Malkin, *Polymer Rheology.* Kymya, Moscow (1977). [English translation by G. V. Vinogradov, Springer-Verlag, Berlin, NY (1981).]
5. R. B. Bird, O. Hassager, R. C. Armstrong and C. F. Curtis, *Dynamics of Polymeric Liquids: Volume II, Kinetic Theory.* John Wiley & Sons, NY (1977).
6. C. D. Han, *Rheology in Polymer Processing.* Adademic Press, NY (1976).
7. VDI-Gesellschaft Kunststofftecknik, *Praktische Rheologie der Kunststoffe.* VDI-Verlag, Duesseldorf (1978).
8. Z. Tadmor and C. G. Gogos, *Principles of Polymer Processing.* John Wiley & Sons, NY (1979).
9. J. L. Throne, *Plastics Process Engineering.* Marcel Dekker, NY (1979).
10. S. Middleman, *Fundamentals of Polymer Processing.* McGraw-Hill, NY (1977).
11. J. R. A. Pearson, *Mechanical Principles of Polymer Melt Processing.* Pergamon Press, London (1966), 2nd Ed. in preparation.
12. R. W. Whorlow, *Rheological Techniques.* Ellis Horwood, distributed by John Wiley & Sons, NY (1979).
13. C. W. Macosko and M. Tirrell, *Rheological Measurements.* University of Minnesota, Dept. of Continuing Education, Short Course Notes (June 1980).
14. K. Walters, *Rheometry.* Chapman & Hall, London (1975).
15. K. Walters, Ed., *Rheometry: Industrial Applications.* John Wiley & Sons, NY (1980).
16. J. L. Leblanc, *Rheologie Experimentale des Polymères a l'Etat Fondu.* Editions CEBEDOC, 2 rue A. Stévart, Liège, Belgium (1974).
17. W. B. Schowalter, *Mechanics of Non-Newtonian Fluids.* Pergamon Press, Elmsford, NY (1978).
18. G. Astarita and G. Marrucci, *Principles of Non-Newtonian Fluid Mechanics.* McGraw-Hill, London (1974).
19. A. S. Lodge, *Elastic Liquids.* Academic Press, NY (1964).
20. A. S. Lodge, *Body Tensor Fields in Continuum Mechanics.* Academic Press, NY (1974).

267

21. C. J. S. Petrie, *Extensional Flows.* Pittman, London (1979).
22. A. S. Lodge and K. Walters, *Rheol. Acta* **14**:573 (1975).
23. B. D. Coleman, H. Markovitz and W. Noll, *Viscometric Flows of Non-Newtonian Fluids.* Springer-Verlag, NY (1966).
24. B. Elbirli and M. T. Shaw, *J. Rheol.* **22**:561 (1978).
25. R. I. Tanner, *A.I.Ch.E. J.* **22**:910 (1976).
26. J. L. White, *Appl. Poly. Symp.* **33**:31 (1978).
27. R. A. Stratton and A. F. Butcher, *J. Polym. Sci. A-2,* **9**:1703 (1971).
28. W. P. Cox and E. H. Merz, *J. Polym. Sci.* **28**:619 (1958).
29. T. T. Tee and J. M. Dealy, *Trans. Soc. Rheol.* **19**:595 (1975).
30. A. C. Pipkin, *Quart. Appl. Math.* **26**:87 (1968).
31. G. V. Vinogradov, *Polymer* **18**:1275 (1977).
32. N. Adams and A. S. Lodge, *Phil. Trans.* **A256**:149 (1964).
33. J. M. Broadbent and A. S. Lodge, *Rheol. Acta* **2**:557 (1971).
34. R. I. Tanner and A. C. Pipkin, *Trans. Soc. Rheol.* **13**:471 (1969).
35. C. D. Han, *A.I.Ch.E. J.* **18**:116 (1972).
36. S. M. Fruh and F. Rodriguez, *A.I.Ch.E. J.* **16**:907 (1970).
37. M. Fujiyama and M. Takayanagi, *Kogyo Kagaku Zasshi* **72**:1163 (1969). (This journal is now called *Nippon Kagaku Kaishi, J. Chem. Soc. Japan, Chem. and Ind. Chem.*)
38. H. M. Laun and J. Meissner, *Rheol. Acta* **19**:60 (1980).
39. S. Middleman, *Trans. Soc. Rheol.* **13**:123 (1969).
40. F. P. La Mantia, B. de Cindio, E. Sorta and D. Acierno, *Rheol. Acta* **18**:369 (1979).
41. A. Pochettino, *Nuovo Cimento* **8**:77 (1914).
42. A. W. Myers and J. A. Faucher, *Trans. Soc. Rheol.* **12**:183 (1968).
43. R. V. McCarthy, *J. Rheol.* **22**:623 (1978).
44. T. L. Smith, J. D. Ferry and F. W. Schremp, *J. Appl. Phys.* **20**:144 (1949).
45. E. Ashare, R. B. Bird and J. A. Lescarboura, *A.I.Ch.E. J.* **11**:910 (1965).
46. R. J. McLachlan, *J. Phys. E.* **9**:391 (1976).
47. F. Ramsteiner, *Rheol. Acta* **15**:427 (1976).
48. K. K. Chee and A. Rudin, *Can. J. Chem. Eng.* **48**:362 (1970).
49. K. K. Chee and A. Rudin, *Rheol. Acta* **16**:635 (1977).
50. A. Rudin, C. K. Ober and K. K. Chee, *Rheol. Acta* **17**:312 (1978).
51. B. Rabinowitsch, *Z. Phys. Chem.* **145**:1 (1929).
52. A. P. Metzger and J. R. Knox, *Trans. Soc. Rheol.* **9**:13 (1965).
53. E. B. Bagley, *J. Appl. Phys.* **28**:624 (1957).
54. R. C. Penwell, R. S. Porter and S. Middleman, *J. Polym. Sci. A-2,* **9**:731 (1971).
55. M. R. Kamal and H. Nyun, *Polym. Eng. Sci.* **20**:109 (1980).
56. W. Philipoff and F. H. Gaskins, *Trans. Soc. Rheol.* **2**:263 (1958).
57. E. B. Bagley, *J. Appl. Phys.* **31**:1126 (1960).
58. C. D. Han and M. Charles, *Trans. Soc. Rheol.* **15**:371 (1971).
59. C. D. Han, *Trans. Soc. Rheol.* **17**:375 (1973).
60. R. Daryanani, H. Janeschitz-Kriegl, R. Van Donselaar and J. Van Dam, *Rheol. Acta* **12**:19 (1973).
61. R. B. Bird, *S.P.E. J.* **11**, *No. 7*:35 (1955).

62. B. Martin, *Int. J. Non-Linear Mech.* **2**:285 (1967).
63. H. W. Cox and C. W. Macosko, *A.I.Ch.E. J.* **20**:785 (1974).
64. N. Galili, Z. Rigbi and R. Takserman-Krozer, *Rheol. Acta* **14**:896 (1975).
65. C. A. Hieber, *Rheol. Acta* **16**:553 (1977).
66. J. E. Gerrard, F. E. Steidler and J. K. Appeldoorn, *Ind. Eng. Chem. Fundam.* **5**:260 (1966).
67. S. Middleman, *The Flow of High Polymers.* Interscience Publishers, NY (1968).
68. H. H. Winter, *Polym. Eng. Sci.* **15**:84 (1975).
69. R. C. Penwell and R. S. Porter, *J. Appl. Polym. Sci.* **13**:2427 (1969).
70. N. Bergem, *Proc. VIIth Int. Congr. Rheol.* Gothenburg (1976), p. 50.
71. Y. Oyanagi, *Appl. Poly. Symp.* **20**:123 (1973).
72. A. Rudin and R.-J. Chang, *J. Appl. Polym. Sci.* **22**:781 (1978).
73. G. V. Vinogradov, N. I. Insarova, B. B. Boiko and E. K. Borisenkova, *Polym. Eng. Sci.* **12**:323 (1972).
74. E. Uhland, *Rheol. Acta* **18**:1 (1979).
75. J. L. White, *Appl. Poly. Symp.* **20**:155 (1973).
76. K. Arisawa and R. S. Porter, *J. Appl. Polym. Sci.* **14**:879 (1970).
77. K. F. Wissbrun and A. C. Zahorchak, *J. Polym. Sci. A-1,* **9**:2093 (1971).
78. S. Negami and R. V. Wargin, *J. Appl. Polym. Sci.* **12**:123 (1968).
79. J. H. Elliott, *Trans. Soc. Rheol.* **12**:573 (1968).
80. E. Hunter and W. G. Oakes, *Brit. Plast.* **17**:94 (1945).
81. J. P. Tordella and R. E. Jolly, *Mod. Plast.* **31**, *No. 2*:146 (1953).
82. R. E. Wiley, *Plast. Technol.* **7**:45 (1961).
83. R. C. Kowalski, *Plast. Eng.* (June 1974), p. 47.
84. H. P. Schreiber and A. Rudin, *J. Appl. Polym. Sci.* **3**:122 (1960).
85. B. Maxwell and A. Jung, *Mod. Plast.* **35**, *No. 3*:174 (November 1957).
86. J. J. Benbow and P. Lamb, *J. Sci. Instrum.* **41**:203 (1964).
87. H. P. Schreiber, *J. Appl. Polym. Sci.* **4**:38 (1960).
88. F. Ramsteiner, *Kunststoffe* **62**:766 (1972).
89. J. P. Tordella, *J. Appl. Phys.* **27**:454 (1956).
90. C. L. Sieglaff, *S.P.E. Trans.* **4**, *No. 2*:129 (1964).
91. R. A. Mendelson, *J. Rheol.* **23**:545 (1979).
92. R. F. Westover, *S.P.E. Trans.* **1**, *No. 1*:14 (1961).
93. R. F. Westover, *Polym. Eng. Sci.* **6**:83 (1966).
94. K. Ito, M. Tsutsui, M. Kasajima and T. Ojima, *Appl. Poly. Symp.* **20**:109 (1973).
95. V. H. Karl, *Angew. Makromol. Chem.* **79**:11 (1979).
96. E. H. Merz and R. E. Colwell, *ASTM Bull.* **232**:63 (1958).
97. J. J. Benbow, *Lab. Practice* **12**:6 (June 1963).
98. D. E. Hanson, *Polym. Eng. Sci.* **9**:405 (1969).
99. A. P. Kiselev, I. F. Kanavets, V. N. Tsventko, A. S. Pereverto and Yu. Boyarski, *Soviet Plast.* (1969), *No. 12*:17, (Translation of *Plast. Massy* **69**, *No. 1*:21.)
100. J. P. Chalifoux and E. A. Meinecke, *J. Polym. Sci. B (Polym. Lett.)* **18**:265 (1980).
101. N. J. Mills, *Rheol. Acta* **8**:226 (1969).

102. E. Langer and U. Werner, *Rheol. Acta* **14**:935 (1975).
103. R. M. Ybarra and R. E. Eckert, *A.I.Ch.E. J* **26**:751 (1980).
104. R. Eswaran, H. Janeschitz-Kriegl and J. Schijf, *Rheol. Acta* **3**:83 (1963).
105. J. L. S. Wales, J. L. den Otter and H. Janeschitz-Kriegl, *Rheol. Acta* **4**:146 (1965).
106. C. D. Han, *J. Appl. Polym. Sci.* **15**:2567 (1971).
107. C. D. Han, *Trans. Soc. Rheol.* **18**:163 (1974).
108. G. Ehrmann, *Kunststoffe* **64**:463 (1974).
109. G. Langer and U. Werner, *Rheol. Acta* **14**:237 (1975).
110. P. W. Springer, R. S. Brodkey and R. E. Lynn, *Polym. Eng. Sci.* **15**:583 (1975).
111. J. L. Leblanc, *Polymer* **17**:235 (1976).
112. J. L. Leblanc, *Proc. VIIth Int. Congr. Rheol.* Gothenburg (1976), p. 278.
113. J. L. Leblanc, *Proc. 3rd Int. IFAC Conf.* Brussels (1976), p. 77.
114. G. Robens and H. H. Winter, *Kunststofftechnik* **13**:61 (1974).
115. M. G. Hansen and J. B. Jansma, in *Rheology 2, Proc. VIIIth Int. Congr. Rheol.* Naples (1980). Published by Plenum Press, NY, p. 193.
116. K. Higashitani and W. G. Pritchard, *Trans. Soc. Rheol.* **16**:687 (1972).
117. D. G. Baird, *Trans. Soc. Rheol.* **19**:147 (1975).
118. D. G. Baird, *J. Appl. Polym. Sci.* **20**:3155 (1976).
119. K. Higashitani and A. S. Lodge, *Trans. Soc. Rheol.* **19**:307 (1975).
120. A. S. Lodge, *Report No. 27.* Rheology Research Center, University of Wisconsin, Madison (1974).
121. A. S. Lodge, U.S. Patent 3,777,549 (1973).
122. A. S. Lodge, *Report No. 60.* Rheology Research Center, University of Wisconsin (March 1980) and Report No. 71 (March 1981).
123. L. de Vargas, Ph.D. Dissertation. University of Wisconsin (1979).
124. C. D. Han and H. J. Yoo, *J. Rheol.* **24**:55 (1980); see also **25**:363, 367 (1981).
125. T.-H. Hou, P. P. Tong and L. de Vargas, *Rheol. Acta* **16**:544 (1977).
126. R. E. Nickell, R. I. Tanner and B. Caswell, *J. Fl. Mech.* **65**:189 (1974).
127. B. A. Whipple and C. T. Hill, *A.I.Ch.E. J.* **24**:664 (1978).
128. M. Gottlieb and R. B. Bird, *Ind. Eng. Chem. Fundam.* **18**:357 (1979).
129. D. V. Boger and M. M. Denn. *J. Non-Newt. Fl. Mech.* **6**:163 (1980).
130. R. I. Tanner, *Trans. Soc. Rheol.* **11**:347 (1967).
131. P. F. Lobo and H. R. Osmers, *Rheol. Acta* **13**:457 (1974).
132. G. Ehrmann, *Rheol. Acta* **15**:8 (1976).
133. S. Okubo and Y. Hori, *J. Rheol.* **24**:275 (1980).
134. H. R. Osmers and P. F. Lobo, *Trans. Soc. Rheol.* **20**:239 (1976).
135. K. Geiger and H. H. Winter, *Rheol. Acta* **17**:264 (1978).
136. K. Geiger, in *Rheology 2, Proc. VIIIth Int. Congr. Rheol.* Naples (1980). Published by Plenum Press. NY, p. 167.
137. T. M. T. Yang and I. M. Krieger, *J. Rheol.* **22**:413 (1978).
138. M. Kepes, *J. Polym. Sci.* **22**:409 (1956).
139. D. J. Highgate and R. W. Whorlow, *Rheol. Acta* **8**:142 (1969).
140. F. D. Dexter, *J. Appl. Phys.* **25**:1124 (1954).
141. R. A. McCord and B. Maxwell, *Mod. Plast.* **39**, No. 1:116 (1961).
142. B. Maxwell and My Nguyen, *Polym. Eng. Sci.* **19**:1140 (1979).

143. C. J. Aloisio, S. Matsuoka and B. Maxwell, *J. Polym. Sci. A-2,* **4**:113 (1966).
144. J. L. den Otter, *Rheol. Acta* **8**:355 (1969).
145. M. Horio, S. Onogi and S. Ogihara, *J. Japan Soc. Testing Materials* **10**:350 (1961).
146. M. Horio, T. Fujii and S. Onogi, *J. Phys. Chem.* **68**:778 (1964).
147. G. V. Vinogradov, Yu. Yanovsky and A. I. Isayev, *J. Polym. Sci. A-2,* **8**:1239 (1970).
148. G. V. Vinogradov, A. I. Isayev and E. V. Katsyutsevich, *J. Appl. Polym. Sci.* **22**:727 (1978).
149. M. Rokudai and T. Fujiki, *Appl. Poly. Symp.* **20**:23 (1973).
150. J. M. Dealy, J. F. Petersen and T. T. Tee, *Rheol. Acta* **12**:550 (1973).
151. V. Semjonow, *Rheol. Acta* **2**:138 (1962).
152. *Ibid.* **4**:133 (1965).
153. F. N. Cogswell, *Plast. Polym.* (February 1973), p. 39.
154. L. Christmann and W. Knappe, *Rheol. Acta* **15**:296 (1976).
155. E. Kuss, *Schmiertechnik und Tribologie* **25**:10 (1978).
156. H. Revesz, *Kunststoffe* **64**:35 (1974).
157. G. Brindley and J. M. Broadbent, *Rheol. Acta* **12**:48 (1973).
158. M. J. Miller and E. B. Christiansen, *A.I.Ch.E. J.* **18**:600 (1972).
159. R. B. Bird and R. M. Turian, *Chem. Eng. Sci.* **17**:331 (1962).
160. G. Heuser and E. Krause, *Rheol. Acta* **18**:553 (1979).
161. D. C. H. Cheng, *Chem. Eng. Sci.* **23**:895 (1968).
162. R. M. Turian, *Ind. Eng. Chem. Fundam.* **11**:361 (1972).
163. J. G. Savins and A. B. Metzner, *Rheol. Acta* **9**:365 (1970).
164. W. M. Kulicke and R. S. Porter, *J. Appl. Polym. Sci.* **23**:953 (1979).
165. R. L. Ballman, L. E. Rademacher and W. H. Farnham, *Appl. Poly. Symp.* **20**:63 (1973).
166. D. J. Paddon and K. Walters, *Rheol. Acta* **18**:565 (1979).
167. J. F. Hutton, *Nature* **200**:646 (1963).
168. J. F. Hutton, *Rheol. Acta* **8**:54 (1969).
169. R. G. King, *Rheol. Acta* **5**:35 (1966).
170. G. V. Vinogradov and A. Ya. Malkin, *Rheol. Acta* **5**:188 (1966).
171. P. T. Gavin and R. W. Whorlow, *J. Appl. Polym. Sci.* **19**:567 (1975).
172. G. Brindley and D. E. Keene, *J. Phys. E.* **7**:934 (1974).
173. D. C. H. Cheng and J. B. Davis, *Report LR42 (CE).* Warren Springs Lab., Stevenage, England.
174. J. Meissner, *J. Appl. Polym. Sci.* **16**:2877 (1972).
175. R. L. Crawley and W. W. Graessley, *Trans. Soc. Rheol.* **21**:19 (1977).
176. A. A. Trapeznikov and A. T. Pylaeva, *Russian J. Phys.-Chem.* **44**:759 (1970). [Translated from *Zhurnal Fiz. Khimii* **44**:1349 (1970).]
177. S. Raha, M. J. Williamson and P. Lamb, *J. Sci. Instrum.* (*J. Phys. E-2*) **1**:1113 (1968).
178. G. V. Vinogradov and I. M. Belkin, *J. Polym. Sci. A,* **3**:917 (1965).
179. K. Komuro, Y. Todani and N. Nagata, *Polym. Letters* **2**:643 (1964).
180. S. Raha, *J. Sci. Instrum.* (*J. Phys. E-2*) **1**:1109 (1968).
181. J. Batchelor, *J. Sci. Instrum.* (*J. Phys. E*) **4**:515 (1971).

182. E. Menefee, *J. Appl. Polym. Sci.* **8**:849 (1964).
183. K. Weissenberg, *Proc. Int. Congr. Rheol.* **2**:114 (1948).
184. J. E. Roberts, *Report 13/52.* Armament Design Establishment, Knockholt, Kent (1952).
185. A. Jobling and J. E. Roberts, *J. Polym. Sci.* **36**:421 (1959).
186. N. J. Mills, A. Nevin and J. McAinsh, *J. Macromol. Sci. (Phys.)* **B4**:863 (1970).
187. N. J. Mills and A. Nevin, *J. Polym. Sci. A-2,* **9**:267 (1971).
188. R. H. Lee, L. G. Jones, K. Pandalai and R. S. Brodkey, *Trans. Soc. Rheol.* **14**:555 (1970).
189. M. Takahashi, T. Masuda and S. Onogi, *Nippon Reoroji Gakkaishi* **5**:72 (1977).
190. S. Ramachandran and E. B. Christiansen, *J. Rheol.* **24**:325 (1980).
191. R. W. Higman, *Rheol. Acta* **12**:533 (1973).
192. J. Meissner, *Rheol. Acta* **14**:201 (1975).
193. E. V. Menezes and W. W. Graessley, *Rheol. Acta* **19**:38 (1980).
194. K. Hayashida and Y. Okuda, *Kagaku Kogaku Rombunshu* **5**:232 (1979).
195. G. H. Piper and J. R. Scott, *J. Sci. Instrum.* **22**:206 (1945).
196. H. A. Pohl and C. G. Gogos, *J. Appl. Polym. Sci.* **5**:67 (1961).
197. K. J. Madonia and C. G. Gogos, *Soc. Plast. Eng., Tech. Pap.* **17**:591 (1971) (29th ANTEC).
198. D. M. Best and S. L. Rosen, *Polym. Eng. Sci.* **8**:116 (1968).
199. E. Broyer and C. W. Macosko, *Soc. Plast. Eng., Tech. Pap.* **21**:343 (1975) (33rd ANTEC).
200. R. A. Jackson and A. Kaye, *Br. J. Appl. Phys.* **17**:1355 (1966).
201. J. F. Petersen, R. Rautenbach and P. Schuemmer, *Rheol. Acta* **14**:968 (1975).
202. K. Sakamoto, N. Ishida and Y. Fukasawa, *J. Polym. Sci.* **6**:1999 (1968).
203. T. Kotaka, M. Kurata and M. Tamura, *J. Appl. Phys.* **30**:1705 (1959).
204. K. Sakamoto and R. S. Porter, *J. Polym. Sci. B,* **8**:177 (1970).
205. D. J. Plazek, *J. Polym. Sci. A-2,* **6**:621 (1968).
206. D. M. Binding and K. Walters, *J. Non-Newt. Fl. Mech.* **1**:277 (1976).
207. M. Mooney, *Ind. Eng. Chem., Anal. Ed.* **6**:147 (1934).
208. J. L. White and N. Tokita, *J. Appl. Polym. Sci.* **9**:1926 (1965).
209. N. Nakajima and E. A. Collins, *Rubber Chem. Technol.* **47**:333 (1974); *48, 69* (1975).
210. N. Nakajima and E. R. Harrell, *Rubber Chem. Technol.* **52**:9 (1979); ibid. p. 962.
211. B. Maxwell and R. P. Chartoff, *Trans. Soc. Rheol.* **9**:41 (1965).
212. B. Maxwell, *Polym. Eng. Sci.* **7**:145 (1967).
213. C. W. Macosko and W. M. Davis, *Rheol. Acta* **13**:814 (1974).
214. L. H. Gross and B. Maxwell, *Trans. Soc. Rheol.* **16**:577 (1972).
215. R. J. J. Jongschaap, K. M. Knapper and J. S. Lopulissa, *Polym. Eng, Sci.* **18**:788 (1978).
216. T. N. G. Abbott and K. Walters, *J. Fl. Mech.* **40**:205 (1970).
217. P. Payvar and R. I. Tanner, *Trans. Soc. Rheol.* **17**:449 (1973).
218. W. M. Davis and C. W. Macosko, *A.I.Ch.E. J.* **20**:600 (1974).
219. F. W. Ahrens and C. Goldstein, *Trans. Soc. Rheol.* **21**:207 (1977).
220. A. Képès, Paper presented at 5th Int. Congr. Rheol. Kyoto (1968). [See also D. H. Kaeble, *J. Appl. Polym. Sci.* **13**:2547 (1969).]

221. F. N. Cogswell, *Trans. Soc. Rheol.* **16**:383 (1972).
222. J. M. Dealy, *J. Non-Newt. Fl. Mech.* **4**:9 (1978).
223. F. N. Cogswell, *Plast. Polym.* **36**:109 (1968).
224. F. N. Cogswell, *Appl. Poly. Symp.* **27**:1 (1975).
225. J. M. Dealy, R. Farber, J. Rhi-Sausi and L. Utracki, *Trans. Soc. Rheol.* **20**:455 (1976).
226. G. V. Vinogradov, V. D. Fikhman and B. V. Radushkevich, *Rheol. Acta* **11**:286 (1972).
227. H. Münstedt, *Rheol. Acta* **14**:1077 (1975).
228. H. M. Laun and H. Münstedt, *Rheol. Acta* **15**:517 (1976).
229. H. M. Laun and H. Münstedt, *Rheol. Acta* **17**:415 (1978).
230. J. Meissner, *Rheol. Acta* **8**:78 (1969).
231. J. Meissner, *Trans. Soc. Rheol.* **16**:405 (1972).
232. J. Meissner, *Pure Appl. Chem.* **42**:553 (1975).
233. H. Münstedt and H. M. Laun, *Rheol. Acta* **18**:492 (1979).
234. T. Raible, A. Demarmels and J. Meissner, *Polym. Bull.* **1**:397 (1979).
235. G. R. Cotten and J. L. Thiele, *Rubber Chem. Technol.* **51**:749 (1978).
236. C. W. Macosko and J. M. Lorntson, *Soc. Plast. Eng. Tech. Pap.* **19**:461 (1973).
237. A. E. Everage and R. L. Ballman, *J. Appl. Polym. Sci.* **20**:1137 (1976).
238. R. W. Connelly, L. J. Garfield and G. H. Pearson, *J. Rheol.* **23**:651 (1979).
239. Y. Ide and J. L. White, *J. Appl. Polym. Sci.* **20**:2511 (1976); **22**:1061 (1978).
240. Y. Ide and J. L. White, *J. Non-Newt. Fl. Mech.* **2**:281 (1977).
241. R. L. Ballman, *Rheol. Acta* **4**:137 (1965).
242. J. F. Stevenson, *A.I.Ch.E. J.* **18**:540 (1972).
243. M. T. Shaw, *Proc VIIth Int. Congr. Rheol.* Gothenburg (1976), p. 304.
244. G. V. Vinogradov, B. V. Radushkevich and V. D. Fikhman, *J. Polym. Sci. A-2*, **8**:1 (1970).
245. P. K. Agrawal, W. K. Lee, J. M. Lorntson, C. I. Richardson, K. F. Wissbrun and A. B. Metzner, *Trans. Soc. Rheol.* **21**:355 (1977).
246. J. Rhi-Sausi and J. M. Dealy, *Polym. Eng. Sci.* **16**:799 (1976).
247. H. Münstedt, *Trans. Soc. Rheol.* **23**:421 (1979).
248. J. Meissner, T. Raible and S. E. Stephenson, *J. Rheol.*, **25**:1 (1981).
249. J. F. Stevenson (Gen. Tire, Akron), Personal communication.
250. Sh. Chatraei and C.W. Macosko, *J. Rheol.*, in press (1981).
251. L. R. G. Treloar, *Trans. Inst. Rubber Ind.* **19**:201 (1944).
252. A. J. De Vries and C. Bonnebat, *Polym. Eng. Sci.* **16**:93 (1976).
253. D. O. Joye, G. W. Poehlein and C. D. Denson, *Trans. Soc. Rheol.* **16**:421 (1972); **17**, 287 (1973).
254. J. M. Maerker and W. R. Schowalter, *Rheol. Acta* **13**:627 (1974).
255. L. R. Schmidt and J. F. Carley, *Polym. Eng. Sci.* **15**:51 (1975).
256. K. C. Hoover and R. W. Tock, *Polym. Eng. Sci.* **16**:82 (1976).
257. C. D. Denson and R. J. Gallo, *Polym. Eng. Sci.* **11**:174 (1971).
258. C. D. Denson and D. C. Hylton, *Polym. Eng. Sci.* **20**:535 (1980).
259. J. M. Dealy, in *Polym. Rheol. and Plast. Process.*, P. L. Clegg *et al.*, Eds. Plastics & Rubber Institute, London (1976), p. 35.
260. J. Rhi-Sausi and J. M. Dealy, *Polym. Eng. Sci.*, **21**:227 (1981).
261. C. D. Denson and D. L. Crady, *J. Appl. Polym. Sci.* **18**:1611 (1974).

262. S. C.-K. Chung and J. F. Stevenson, *Rheol. Acta* **14**:832 (1975).
263. J. F. Stevenson, S. C.-K. Chung and J. T. Jenkins, *Trans. Soc. Rheol.* **19**:397 (1975).
264. R. G. Foltz, K. K. Wang and J. F. Stevenson, *J. Non-Newt. Fl. Mech.* **3**:347 (1978).
265. A. H. Hoffman and W. G. Gottenburg, *Trans. Soc. Rheol.* **17**:465 (1973).
266. J. M. Dealy and T. K. P. Vu, *J. Non-Newt. Fl. Mech.* **3**:127 (1977/78).
267. P. K. Freakley and W. Y. Wan Idris, *Rubber Chem. Technol.* **52**(*1*):134 (1979).
268. J. E. Goodrich and R. S. Porter, *Polym. Eng. Sci.* **7**:45 (1967).
269. L. L. Blyler and J. H. Daane, *Polym. Eng. Sci.* **7**:178 (1967).
270. G. C. N. Lee and J. R. Purdon, *Polym. Eng. Sci.* **9**:360 (1969).
271. M. G. Rogers, *Ind. Eng. Chem. Process Des. Dev.* **9**:49 (1970).
272. L. P. Yasenchak, *Mod. Plast.* (December 1973).
273. L. W. Mentovay and L. P. Yasenchak, *Plast. Des. Process.* (March 1973).
274. J. Gavis and M. Modan, *Phys. Fl.* **10**:487 (1967).
275. R. I. Tanner, *Appl. Poly. Symp.* **20**:201 (1973).
276. K. Reddy and R. I. Tanner, *J. Rheol.* **22**:661 (1978).
277. L. L. Chapoy, *Rheol. Acta* **8**:497 (1969).
278. E. B. Bagley, S. H. Storey and D. C. West, *J. Appl. Polym. Sci.* **7**:1661 (1963).
279. F. N. Cogswell, *Plast. Polym.* (December 1970), p. 391.
280. E. B. Bagley and H. J. Duffy, *Trans. Soc. Rheol.* **14**:545 (1970).
281. Y. Mori and K. Funatsu, *Appl. Poly. Symp.* **20**:209 (1973).
282. R. Racin and D. C. Bogue, *J. Rheol.* **23**:263 (1979).
283. R. A. Mendelson, F. L. Finger and E. B. Bagley, *J. Polym. Sci. C,* **35**:177 (1971).
284. R. I. Tanner, *J. Polym. Sci. A-2,* **8**:2067 (1970).
285. J. Vlachopoulos, M. Horie and S. Lidorikis, *Trans. Soc. Rheol.* **16**:669 (1972).
286. I. Pliskin, *Rubber Chem. Technol.* **46**:1218 (1973).
287. J. L. White and J. F. Roman, *J. Appl. Polym. Sci.* **20**:1005 (1976).
288. L. A. Utracki, Z. Bakerdjian and M. R. Kamal, *J. Appl. Polym. Sci.* **19**:481 (1975).
289. J. M. Dealy, A. Garcia-Rejon and M. R. Kamal, *Can. J. Chem. Eng.* **55**:651 (1977).
290. J. M. Dealy and A. Garcia-Rejon, *Proc. VIIIth Int. Congr. Rheol.* **3**:63. Plenum Press, NY (1980).
291. P. J. Leider and R. B. Bird, *Ind. Eng. Chem. Fundam.* **13**:336 (1974).
292. G. Brindley, J. M. Davies and K. Walters, *J. Non-Newt. Fl. Mech.* **1**:19 (1976).
293. M. T. Shaw, *Polym. Eng. Sci.* **17**:266 (1977).
294. R. J. Grimm, *A.I.Ch.E. J.* **24**:427 (1978).
295. G. J. Dienes and H. F. Klemm, *J. Appl. Phys.* **17**:458 (1946).
296. A. N. Gent, *Br. J. Appl. Phys.* **11**:85 (1960).
297. T. Kataoka, T. Kitano and T. Nishimura, *Rheol. Acta* **17**:626 (1978).
298. S. Oka and S. Ogawa, *Zairo Shiken* **9**:321 (1960); **12**:324 (1963).
299. C. J. Bartlett, *Soc. Plast. Eng., Tech. Pap.* **24**:638 (1978) (36th ANTEC).
300. D. P. Bloechle, *Ibid.,* p. 641.
301. A. B. Metzner and M. M. Denn, Univ. of Delaware, Personal communication.
302. G. Marrucci and R. E. Murch, *Ind. Eng. Chem. Fundam.* **9**:498 (1970).

303. J. L. White and A. Kondo, *J. Non-Newt. Fl. Mech.* **3**:41 (1977).
304. A. E. Everage and R. L. Ballman, *J. Appl. Polym. Sci.* **18**:933 (1974).
305. H. P. Hürlimann and W. Knappe, *Rheol. Acta* **11**:292 (1972).
306. V. I. Brizitsky, G. V. Vinogradov, A. I. Isayev and Y. Y. Podolsky, *J. Appl. Polym. Sci.* **22**:751 (1978).
307. F. N. Cogswell, *J. Non-Newt. Fl. Mech.* **4**:23 (1978).
308. F. N. Cogswell, *Polym. Eng. Sci.* **12**:64 (1972).
309. R. N. Shroff, L. V. Cancio and M. Shida, *Trans. Soc. Rheol.* **21**:429 (1977).
310. M. T. Shaw, *J. Appl. Polym. Sci.* **19**:2811 (1975).
311. A. E. Everage and R. L. Ballman, *Nature* **273**:213 (1978).
312. H. H. Winter, C. W. Macosko and K. E. Bennett, *Rheol. Acta* **18**:323 (1979).
313. E. D. Johnson and S. Middleman, *Polym. Eng. Sci.* **18**:963 (1978).
314. G. H. Pearson and S. Middleman, *A.I.Ch.E. J.* **23**:714 (1977).
315. H. Münstedt and S. Middleman, *J. Rheol.,* **25**:29 (1981).
316. J. W. Hill and J. A. Cuculo, *J. Macromol. Sci., Rev. Macromol. Chem.* **C14**:107 (1976).
317. W. F. Busse, *J. Polym. Sci. A-2,* **5**:1249 (1967).
318. F. N. Cogswell, *Rheol. Acta* **8**:187 (1969).
319. J. Meissner, *Rheol. Acta* **10**:230 (1971).
320. K. Wissbrun, *Polym. Eng. Sci.* **13**:342 (1973).
321. M. Swerdlow, F. N. Cogswell and N. Krul, *Plast. Rubber Process.,* in press (1980).
322. G. M. Fehn, *J. Polym. Sci. A-1,* **6**:247 (1968).
323. D. Acierno, J. N. Dalton, J. M. Rodriguez and J. L. White, *J. Appl. Polym. Sci.* **15**:2395 (1971).
324. C. D. Han and R. R. Lamonte, *Trans. Soc. Rheol.* **16**:447 (1972).
325. E. Deprez and W. J. Bontinck, in *Polym. Rheol. and Plast. Process.,* P. L. Clegg *et al.,* Eds. Plastics & Rubber Institute, London (1975), p. 274.
326. R. K. Bayer, H. Schreiner and W. Ruland, *Rheol. Acta* **17**:28 (1978).
327. R. K. Bayer, *Rheol. Acta* **18**:25 (1979).
328. J. A. Spearot and A. B. Metzner, *Trans. Soc. Rheol.* **16**:495 (1972).
329. I.-J. Chen, G. E. Hagler, L. E. Abbott, D. C. Bogue and J. L. White, *Trans. Soc. Rheol.* **16**:473 (1972).
330. C. D. Han and J. Y. Park, *J. Appl. Polym. Sci.* **19**:3257 (1975).
331. K. Iwakura, M. Yoshinari and T. Fujimura, *J. Soc. Rheol. Japan* **3**:134 (1975).
332. R. Farber and J. M. Dealy, *Polym. Eng. Sci.* **14**:435 (1974).
333. J. C. Miller, *Trans. Soc. Rheol.* **19**:341 (1975).
334. J. M. Dealy and A. Garcia-Rejon, Paper presented to Society of Rheology. Boston (October 1979).
335. H. Münstedt, *Eng. Found. Conf.* Asilomar (January 1980).
336. M. K. Black, *Plast. Des. Process.* (December 1963).
337. P. L. Shah, *S.P.E. J.* **27**, No. 1 (January 1971).
338. L. L. Blyler, Jr., *Polym. Eng. Sci.* **14**:806 (1974).
339. O. F. Noel III and J. F. Carley, *Polym. Eng. Sci.* **15**:117 (1975).
340. M. P. van der Wielen, *Polym. Eng. Sci.* **15**:102 (1975).
341. J. Warmuth, *Plastverarbeiter* **1/2** (1973).
342. H. Wesche, *Kautschuk und Gummi Kunstoffe* **31**:495 (1978).

343. H.-E. Toussaint, W. N. Unger and H. O. Schäfer, *Ibid.* **25**:155 (1972).
344. J. F. Agassant, J. P. Villemaire and P. Avenas, Paper presented at Europlastique (1978) in Paris.
345. R. O. Welty, *Soc. Plast. Eng., Tech. Pap.* **11** (1965).
346. F. H. McGinnis and C. D. Han, *Soc. Plast. Eng., Tech. Pap.* **24**:227 (1978) (36th ANTEC).
347. C. Macosko and J. M. Starita, *S.P.E. J.* **27**:38 (November 1971).
348. N. Nakajima, D. W. Ward and E. A. Collins, *J. Appl. Polym. Sci.* **20**:1187 (1976).
349. R. G. Gough and I.-J. Chen, *Polym. Eng. Sci.* **17**:764 (1977).
350. N. Nakajima and E. A. Collins, *Trans. Soc. Rheol.* **20**:1 (1976).
351. A. Ram, *Polym. Eng. Sci.* **17**:793 (1977).
352. J. M. Starita, in *Rheology 2, Proc. VIIIth Int. Congr. Rheol.* Naples (1980). Published by Plenum Press, NY, p. 229.
353. C. J. Drislane, J. P. De Nicola, W. M. Wareham and R. I. Tanner, *Rheol. Acta* **13**:4 (1974).
354. W. Heinz and A. Rothenpieler, *Proc. VIIth Int. Congr. Rheol.* Gothenburg (1976), p. 458.
355. M. Seeger and W. Heinz, in *Rheology 2, Proc. VIIIth Int. Congr. Rheol.* Naples (1980). Published by Plenum Press, NY, p. 173.
356. V. M. Lobe and C. W. Macosko, Paper presented to Society of Rheology. Boston, November 1979.
357. W. D. Arendt, *Plast. Eng.* (September 1979), p. 46.
358. R. P. Petrich, *Mod. Plast.* (August 1972).
359. E. A. Collins, A. P. Metzger and R. R. Furgason, *Polym. Eng. Sci.* **16**:240 (1976).
360. I. Patterson and J. Noah, *Soc. Plast. Eng., Tech. Pap.* **23**:340 (1977) (35th ANTEC).
361. E. O. Allen II and R. F. Williams, Jr., *Polym. Eng. Sci.* **12**:353 (1972).
362. J. E. Goodrich, *Polym. Eng. Sci.* **10**:215 (1970).
363. P. E. Willard, *Polym. Eng. Sci.* **14**:273 (1974).
364. ———, *Chemical and Plastics Industry Bibliography,* 10th Ed. C. W. Brabender Instruments, Hackensack, NJ (1976).
365. H. M. Kromer, *Introduction to Torque Rheometry* Haake, Saddle Brook, NJ (1978).
366. K. T. Paul, *RAPRA Bull.* (February 1972), p. 29.
367. P. J. Heinle and M. A. Rodgers, *S.P.E. J.* **25**:6, 56 (1969).
368. R. P. White, Jr., *Polym. Eng. Sci.* **14**:50 (1974).
369. F. G. Mussatti and C. W. Macosko, *Polym. Eng. Sci.* **13**:263 (1973).
370. F. R. Volgstadt and C. L. Sieglaff, *Polym. Eng. Sci.* **14**:143 (1974).
371. E. B. Richter and C. W. Macosko, *Polym. Eng. Sci.* **20**:921 (1980).
372. M. T. Shaw, S. Burket and D. W. Sundstrom, *Rev. Sci. Instrum.* **49**:1597 (1978).
373. J. M. Dealy and W. K. W. Tsang, *J. Appl. Polym. Sci.,* **26**:1149 (1981).
374. B. Maxwell and J. Plumeri, *S.P.E. Tech. Pap.* **26**:282 (1980).
375. M. N. Nguyen and B. Maxwell, *Polym. Eng. Sci.* **20**:972 (1980).

APPENDICES

Appendix A
Units and Conversions

QUANTITY AND SOME EQUIVALENCES	TO CONVERT FROM	TO	MULTIPLY BY
Length			
1 m = 100 cm	inch (in.)	meter (m)	2.54×10^{-2}
	inch (in.)	centimeter (cm)	2.54
	foot (ft)	meter (m)	0.3048
	centimeter (cm)	meter (m)	0.010
	meter (m)	foot (ft)	3.281
Volume			
$1\ m^3 = 10^3$ liter	$foot^3\ (ft^3)$	$meter^3\ (m^3)$	0.02832
$1\ m^3 = 10^6\ cm^3$	$inch^3\ (in.^3)$	$meter^3\ (m^3)$	1.639×10^{-5}
	liter (l)	$meter^3\ (m^3)$	10^{-3}
	$centimeter^3\ (cm^3)$	$meter^3\ (m^3)$	10^{-6}
	$meter^3\ (m^3)$	$foot^3\ (ft^3)$	35.31
Force (F)			
$N \equiv kg \cdot m/s^2$	pound force (lbf)	newton (N)	4.448
$1\ N = 10^5$ dyne	dyne	newton (N)	10^{-5}
$dyne \equiv g \cdot cm/s^2$	kilogram force	newton (N)	9.807
Pressure, Stress			
$Pa \equiv N/m^2$	$lbf/in.^2$ (psi)	pascal (Pa)	6.895×10^3
$Pa = 10\ dyne/cm^2$	$dyne/cm^2$	pascal (Pa)	0.10
$bar \equiv 10^5\ Pa$	$kg\ (force)/cm^2$	bar	0.9807
$bar = 10^6\ dyne/cm^2$	$kg\ (force)/m^2$	$dyne/cm^2$	98.07
	$kg\ (force)/m^2$	pascal (Pa)	9.807
	pascal (Pa)	bar	10^{-5}
	bar	$lbf/in.^2$ (psi)	14.51
Viscosity (η)			
$poise \equiv \dfrac{dyne \cdot s}{cm^2} = \dfrac{g}{cm \cdot s}$	poise	$Pa \cdot s$	0.100
	centipoise	$Pa \cdot s$	10^{-3}
$Pa \cdot s \equiv N \cdot s/m^2$	$Pa \cdot s$	poise	10.0
1 poise = 100 centipoise			
$1\ Pa \cdot s = 10$ poise			
1 centipoise = 1 $mPa \cdot s$			

QUANTITY AND SOME EQUIVALENCES	TO CONVERT FROM	TO	MULTIPLY BY
Torque (M)			
$M = F \times R$	meter·g (force)	m·N	0.00981
$1 \ m·N = 10^7 \ cm·dyne$	m·kg (force)	m·N	9.81
	ft·lbf	m·N	1.356
	cm·dyne	m·N	10^{-7}
	in·lbf	m·N	0.113
Rotational Speed (Ω) and Frequency (ω)			
Hertz (Hz) $\equiv$ cycle/sec.	revolutions/min. (RPM)	radian/sec.	0.1047
1 cycle = 2π radian (rad)	cycle/sec. (Hz)	radian/sec.	6.283
1 revolution = 2π radian (rad)	rad/sec.	cycle/sec. or revolutions/sec.	0.1592
Power or Dissipation Rate			
watt (W) $\equiv$ J/s	Btu/hr	watt (W)	0.2931
joule (J) = N·m	horsepower	watt (W)	745.7
Thermal Conductivity (K)			
$\dfrac{\text{watt}}{\text{meter·Kelvin}} \left(\dfrac{W}{m·K} \right)$	Btu/ft-hr-°F	W/m·K	1.731
$\left(\dfrac{W}{m·K} \right) = \left(\dfrac{N}{s·K} \right) = \left(\dfrac{kg·m}{s^3·K} \right)$	g·cm/s³·K	W/m·K	10^{-5}

Appendix B
Manufacturers and Their Agents

COMPANY NAME AND PRODUCTS	ADDRESSES
Brabender Plasti-Coder and accessories EstrusioGraph Rheotron	C. W. Brabender Instruments, Inc. P. O. Box 2127 So. Hackensack, NJ 07606 Brabender OHG Postfach 350 162 D-4100 Duisberg 1 Fed. Republic of Germany

Brookfield
 	
Synchro-Lectric Viscometer
Thermosel System
  Agent:
  Testing Machines. Inc. (*q.v*)
	Brookfield Engineering Laboratories, Inc.
	240 Cushing Street
	Stoughton, MA

Bruss Torque Measuring Drives (Process Viscometers)	H. K. Bruss Rheo-Verfahrenstechnik GmbH Postfach 1166 6232 Bad Soden Fed. Republic of Germany
Burrell Burrell-Severs Extrusion Rheometer	Burrell Corporation 2223 Fifth Avenue Pittsburgh, PA 15219
CEAST Melt Index Apparatus	CEAST S.p.A. Via Asinari di Bernezzo, 70 10146 Torino Italy

U.S. Agent:
Intermetco
300 Concord Road
Wayland, MA 01778

Contraves

Balance Rheometer

    U.S. Agent:
    Olkon Corporation
    999 Bedford Street
    Stamford, CT 06905

Contraves AG Zurich
Schaffhauserstrasse 580
8052 Zurich
Switzerland

Custom Scientific

Melt Index Apparatus
Melt Elasticity Tester
Orthogonal Rheometer

Custom Scientific Instruments, Inc.
P. O. Box A
Whippany, NJ 07981

Davenport

Cone and Plate Viscometer
Extrusion Rheometer
High-Shear Viscometer
Melt-Flow Indexer

    U.S. Agent:
    Testing Machines, Inc. (*q.v.*)

Davenport (London), Limited
Tewin Road
Welwyn Gardon City
Herts AL7 1AQ
England

Deimos

Piezoelectric sensing head and
    charge amplifier
    (accessories for Rheogoniometer)

Deimos, Limited
Simmonds Road
Wincheap Industrial Estate
Canterbury
England

Dynisco

Pressure transducers for molten
    plastics

Dynisco
20 Southwest Park
Westwood, MA 02090

Du Pont

Parallel Plate Rheometer
(Thermomechanical Analyzer)

DuPont Company, Instrument
  Products
Scientific & Process Division
Wilmington, DE 19898

Karl Frank

Melt Index Tester
Mooney Viscometer

Karl Frank GmbH
Postfach 1320
D-6940 Weinheim
Fed. Republic of Germany

GCA/Precision Scientific

(Scott Testers)
Automatic Mooney Viscometer

GCA Corporation
Precision Scientific Group
3737 West Cortland Street
Chicago, IL 60647

Göttfert

High Pressure Capillary Rheometer
Kontirheograph
Minex
Rheograph
Rheostrain
Rheotens
Side-Stream Capillary Rheometer
Torsiograph

Göttfert Werkstoff-Prüfmaschinen
  GmbH
Postfach 1220
6967 Buchen
Fed. Republic of Germany

U.S. Agent:
AUTOMATIK
Maschinenbau of USA, Ltd.
460 Summer Street
Stamford, CT 06901

Haake

Rheocord
Rheomix, Rheolex, Rheomex
Consistometer

Haake, Inc.
244 Saddle River Road
Saddle Brook, NJ 07662

Gebrüder Haake
Dieselstrasse 4
D75 Karlsruhe 41
Fed. Republic of Germany

Hampden

RAPRA Variable Torque
  Rheometer

Hampton Test Equipment, Limited
Rothersthorpe Avenue
Northampton , NN4 9JH
England

Imass

Agent for the Sieglaff-McKelvey
  Rheometer made by Tinius-
  Olsen and for Dynalizer made by
  Dynastatics

Imass, Inc.
P. O. Box 134
Accord (Hingham), MA 02018
U.S.A.

Instron

Capillary Rheometers
Rotational Rheometer
General purpose testing machines

Instron Corporation
100 Royall Street
Canton, MA 02021

Instron, Limited
Coronation Road
High Wycombe
Bucks HP12 3SY
England

Instron GmbH
Frankenstrasse
D-8757 Karlstein 2
Fed. Republic of Germany

Iwamoto

Melt Viscometer
Mini-Viscometer
3-D Rheometer
Rheometer Almighty

Iwamoto Seisakusho Co., Ltd.
354 Hazukashi Furukawacho
Kyoto, Japan

Kayeness

Extrusion Plastometer
Computer Controlled Melt
  Rheometer

Kayeness, Inc.
RD 3, Box 30
Honeybrook, PA 19344

Lebow

Load Cells
Torque Transducers

Lebow Associates, Inc.
P. O. Box 1089
Troy, MI 48099

MTS

Servohydraulic Testing Machines

MTS Systems Corporation
P. O. Box 24012
Minneapolis, MN 55424

Metrilec

Rheoplast

Metrilec SARL
20, rue Michal
75013 Paris
France

Monsanto

Automatic Capillary Rheometer
Automatic Die Swell Detector
Melt Plastometer
Processability Tester

Monsanto Industrial Chemicals
947 West Waterloo Road
Akron, OH 44314

Monsanto Technical Center
Parc Industriel
Rue Laid Burniat
B-1348 Louvain-la-Neuve
Belgium

Petronic

Viscoelastic analyzer

Integrated Petronic Systems, Ltd.
120 Woodgrange Road
Forest Gate
London E7 OEW
England

Polymer Products and Services

Molds for cone plate and extensional
  rheometer samples

Polymer Products and Services
P. O. Box 13927
Gainsville, FL 32604

RML

Deer Rheometer

Rheometer Marketing, Limited
Crown House
Amley Road
Leeds LS12 2EJ
England

Ray-Ran

Mk 1 Utility Melt Indexer
Mk 2 Digital Melt Indexer

U.S. Agent: TMI (*q.v.*)

Ray-Ran Engineering
Dawson House
Bennetts Road
Keresley CV7 8HY
England

Rheometrics

Dynamic Spectrometer
Extensional Rheometer
Mechanical Spectrometer
System Four
Visco-Elastic Tester

Rheometrics, Inc.
2438 U.S. Highway No. 22
Union, NJ 07083

Rheometrics GmbH
Arabella Center
Lyoner Strasse 44-48
D6000 Frankfurt a.M. 71
Fed. Republic of Germany

Rodem

Dynamic Viscometer

Instrumentfabriek Rodem
Rembrandtlaan 5a
3723 BG Bilthoven
Netherlands

Sangamo

Weissenberg Rheogoniometer

Sangamo Schlumberger
Rheology Division
North Bersted
Bognor Regis
Sussex. PO22 9BS
England, U.K.

1875 Grand Island, Blvd.
Grand Island, NY 14072

Seiscor

Continuous Melt Rheometer
Continuous Capillary Rheometer
Solution and Particle Sampling
    Systems
Seiscor/Han Rheometer

Seiscor Division
P. O. Box 1590
Tulsa, OK 74102

Shimadzu

Koka Flow Tester

Shimadzu Seisakusho, Limited
14-5, Uchikanda 1-chome
Chiyoda-Ku
Tokyo 101
Japan

Slocumb

Extrusion Plastomer

F. F. Slocumb Corporation
P. O. Box 1591
Wilmington, DE 19899

Solartron

Analyzer for oscillatory
    shear signals

Solartron Electronic Group, Limited
Farnborough
Hants GU14 7PW
England

TMI

Agents for Davenport and Wallace

Testing Machines, Inc.
400 Bayview Avenue
Amityville, NY 11701

Tinius-Olsen

Extrusion Plastometer
Olsen Parallel Plate Plastometer
Olsen-Bakelite Flow Tester
Sieglaff-McKelvey Rheometer
(Marketed by Imass, *q.v.*)

Tinius-Olsen Testing Machine Co.,
    Inc.
Easton Road
Willow Grove, PA 19090

Toyoseiki

Melt Indexer
Automatic Melt Indexer
Labo Plastomill
Strograph (testing machines)
Westover Type Rheometer
Melt Strength Device

Toyo Seiki Seisaku-sho, Limited
15-4, 5-Chome
Takinogawa, Kita-Ku
Tokyo 114
Japan

Wallace

Melt Flow Index Tester

    U.S. Agent:
    Testing Machines, Inc. (*q.v.*)

H. W. Wallace & Co., Ltd.
St. James Road
Croydon
England

Zwick

Capillary Plastometer

Zwick GmbH & Co.
Postfach 4350
D-7900 Ulm
Fed. Republic of Germany

Appendix C
Price Ranges for Commercial Instruments

Price information is given in terms of approximate ranges as of mid-1981. The manufacturer or its agent should be consulted for accurate, current prices, as these have been changing rapidly in recent years due to technological innovations and inflation.

Letter Code	Price Range (1981) in U.S. Dollars
A	Under $1,000
B	1,000–2,000
C	2,000–3,000
D	3,000–4,000
E	4,000–5,000
F	5,000–10,000
G	10,000–15,000
H	15,000–20,000
I	20,000–25,000
J	25,000–30,000
K	30,000–50,000
L	50,000–100,000
M	100,000–150,000
N	Over $150,000

NOMENCLATURE

a_i Coefficients in velocity distribution (Equation 2-14)
A Cross-sectional area
A_c Area of cylinder (Equation 9-4)
A_0 Initial value of cross-sectional area of specimen
A_p Area of piston

b	Temperature sensitivity of viscosity (Equation 3-7) or Rabinowitsch correction (Equation 4-19b)
B	Capillary extrudate swell ratio (Equation 7-2)
d	Diameter of circle shown in Figure 6-3 or offset of disks in ERD flow
d_0	Initial value of d
D	Diameter
D_0	Capillary diameter
D_e^0	Steady state elongational compliance
$D(t)$	Extensional creep compliance
e	End correction (Equation 4-28)
$E(s)$	Extensional relaxation modulus
F	Force
F_d	Driving force for capillary flow
G	Constant shear modulus (Equation 2-55)
$G(t)$	Shear stress relaxation function (Equation 2-32)
G'	Storage modulus
G''	Loss modulus
h	Gap spacing or sheet height in Figure 6-2
H	Cylinder displacement
J	Shear creep compliance (Equation 2-50)
J_e^0	Steady state shear compliance (Equation 2-51)
k	Power law coefficient (Equation 2-10) or thermal conductivity
L	Sample length or capillary length
L_0	Initial length of sample
M	Torque
M_0	Amplitude of oscillating torque
n	Power law index (Equation 2-10) or viscous end correction (Equation 4-30)
N_1	First normal stress difference (Equation 2-2)
N_2	Second normal stress difference (Equation 2-3)
N^+	Normal stress growth function
N^-	Normal stress relaxation function; cessation of steady shear
P	Pressure
P_a	Ambient pressure
P_c	Gas pressure in cylinder (Equation 9-4) or capillary pressure (Equation 4-47)
P_d	Driving pressure in capillary flow
P_H	Pressure hole error
P_m	Pressure measured in pressure hole
P_w	Wall pressure
P_{ec}	Exit pressure for capillary flow
P_{es}	Exit pressure for slit flow

Q	Volumetric flow rate
r	Radial coordinate in cylindrical coordinate system
r_i	Inner radius
r_0	Outer radius
R	Radius of sample, fixture or capillary or radius of curvature
R_b	Barrel radius or bubble radius
s	$t - t'$
S_r	Recoverable shear (Equation 2-54)
t	Time
t'	Point in time between $-\infty$ and t
T	Temperature
T_w	Wall temperature
v_i	Velocity component
v_R	Velocity component in radial direction (spherical coordinates)
V	Velocity of surface
V_p	Velocity of plunger
V_w	Velocity of wall
W	Width of slit
x_i	Rectangular coordinate
X	Displacement of piston
y	Distance from wall or distance defined in Figure 6-3
z	Axial coordinates in cylindrical coordinates
Z	Distance between rotary clamps or spindles
Z_1	Distance from fixed end of sample to rotary clamp
α	Angle defined in Figure 6-2 or surface tension
β	Angle defined in Figure 6-3 or factor in Equation 4-38
γ	Shear strain (Equations 1-2, 1-3)
γ_0	Fixed amount of strain or strain amplitude
$\gamma_r(t)$	Shear strain recovery function (Equation 2-53)
$\dot{\gamma}$	Shear rate (Equation 1-5)
$\dot{\gamma}_0$	Constant shear rate or shear rate amplitude
$\dot{\gamma}_A$	Apparent shear rate at the wall
$\dot{\gamma}_m$	Mean shear rate
$\dot{\gamma}_R$	Shear rate at $r = R$
$\dot{\gamma}_w$	Shear rate at the wall
δ	Mechanical loss angle or thickness of inflated sheet
Δ	Difference or change in a quantity
ϵ	Hencky strain (Equation 1-11)
ϵ_b	Biaxial strain
ϵ_0	Fixed amount of Hencky strain (creep test)
ϵ_r	Recoverable extensional strain
$\dot{\epsilon}$	Extensional strain rate (Equation 2-17)

$\dot{\epsilon}_b$	Biaxial strain rate
$\dot{\epsilon}_0$	Constant extensional strain rate
η	Viscosity (Equation 1-9)
η^+	Stress growth function (constant strain rate)
η^-	Stress relaxation function (cessation of steady straining)
η'	Dynamic viscosity
η''	Imaginary component of complex viscosity
η^*	Complex viscosity
η_0	Low shear rate limiting value of viscosity
η_b	Biaxial extensional viscosity (Equation 2-23)
η_p	Planar extensional viscosity
η_T	Extensional (Trouton) viscosity (Equation 2-19)
$\bar{\eta}_T$	Strain averaged value of η_T
η_{EB}	Apparent extensional viscosity (Equation 7-35)
η_{EC}	Apparent extensional viscosity (Equation 7-19)
η_{ES}	Apparent extensional viscosity (Equation 7-40)
θ	Angular coordinate in azimuthal direction; spherical coordinates, or angular coordinate in cylindrical system
θ_0	Cone angle
λ	Stretch ratio
π	3.14159 or isotropic contribution to total stress
ρ	Density
σ_{ij}	Component of total stress
σ_b	Net biaxial stretching stress (Equation 6-27)
σ_e	Net stretching stress (Equation 6-7)
τ_{ij}	Component of the extra stress (Equation 1-8)
τ	Shear stress in simple shear
τ_i	Shear stress at wall of inner cylinder
τ_m	Mean shear stress
τ_0	Fixed shear stress or shear stress amplitude
τ_w	Shear stress at the wall
ϕ	Angular displacement in rotational flow or angular coordinate about axis of symmetry in spherical coordinates
ϕ_0	Displacement amplitude for rotational oscillation
Ψ_1	First normal stress difference coefficient
Ψ_2	Second normal stress difference coefficient
ω	Frequency (rad/sec.)
Ω	Rotational speed (rad/sec.)

Author Index

Manufacturer Index

Subject Index

American Society for Testing and Materials
  (see ASTM)
Annular flow, 114, 115
Apparent extensional viscosity:
  bubble collapse, 184
  capillary entrance flow, 180
  film blowing, 189
  melt drawing, 188
Apparent wall shear rate:
  in a capillary, 77, 200, 204
  in a slit, 105
Assumptions, 2
ASTM tests:
  D569, 96, 97
  D926, 177
  D1238, 94, 201, 210, 224
  D1647, 141
  D2396, 257
  D2538, 257
Automation of rheometers, 196, 204, 213–215

Biaxial extension:
  definition, 38, 39
  experimental methods, 159–166
Biaxial extensional viscosity, 38
Bibliographies, 21
Biconical rheometer, 138
Birefringence, flow, 103
Books on melt rheology, 20, 21
Bubble collapse, 183–185
Bubble inflation (see sheet inflation)

Calibration of rheometers, 66, 67
Capillary flow:
  analysis of data, 92
  basic equations, 75–80
  controllability of, 75
  degradation in, 91, 92
  end effects, 80–85
  entrance flow, 179–181
  flow instability, 90, 91

pressure effects, 89, 90
temperature distribution, 85–89
viscous heating, 85–89
(see also driving pressure)
Capillary rheometers
  extruder fed, 102, 204
  gas pressure driven, 97–100, 208–214
  high pressure, 100, 214
  hydraulic drive, 103, 215–218
  motor driven, 101, 102, 218–223
  weight driven, 93–97, 202–208
  (see also melt indexers)
Cauchy strain, 4, 5
Cessation of steady shear, 14, 45, 46
CIL flow index, 99, 225
CIL viscometer, 97, 98
  commercial versions, 210, 211
Classification of polymeric materials, 19
Cleaning of rheometer, 66
Complex modulus, 51
Complex viscosity, 51
Compliance:
  shear creep, 47, 48
  extensional creep, 54, 55
  of rheometer, 133, 136, 146
Cone and plate flow, 126
  basic equations, 127–129
  sources of error, 130–133
Cone and plate rheometers, 134–137
  molds for, 234
  (see also rotational rheometers)
Cone angle, selection of, 133, 134
Concentric cylinder flow, 116
  basic equations, 117–120
  sources of error, 121–122
Concentric cylinder rheometers, 123–126
  high pressure models, 125, 126
  (see also rotational rheometers)
Constitutive equation, 10, 11, 27
Continuous property measurement, 190, 204, 222–228
Continuum assumption, 2